Also by Gary Romano
Why I Farm: Risking It All for a Life on the Land
A Hiker's Guide to Wild Edible Plants of San Luis Obispo County

For more about Sierra Valley Farms and Gary's market
and event schedule, go to sierravalleyfarms.com.

JULY & WINTER

JULY & WINTER
Growing Food in the Sierra

Gary Romano

Bona Fide Books
Tahoe Paradise, CA

Copyright © 2016 by Gary Romano.

All rights reserved. No portion of this work may be reproduced or transmitted in any form or by any means, electronic or mechanical, including photocopying and recording, or by any information storage or retrieval system, without permission in writing from Bona Fide Books.

ISBN 978-1-936511-12-9
2016942240

Cover Design: Ponderosa Pine Design
Illustrations: Coco Foy
Copy Editor: Pam Suwinsky
Layout: KristenSchwartz.com
Index: www.mlmindexing.com
Editorial Intern: Meghan Robins
Printing and Binding: Bang Printing
Special thanks to Roger Freeburg for his photos.

Orders, inquiries, and correspondence should be addressed to:
Bona Fide Books
PO Box 550278, South Lake Tahoe, CA 96155
(530) 545-1373
www.bonafidebooks.com

"The single greatest lesson the garden teaches is that our relationship to the planet need not be zero-sum, and that as long as the sun still shines and people still can plan and plant, think and do, we can, if we bother to try, find ways to provide for ourselves without diminishing the world."

Michael Pollan
The Omnivore's Dilemma: A Natural History of Four Meals

CONTENTS

INTRODUCTION	xiii
How to Use This Book	xvi
🌰 Foothill Region	xvii
🌲 Sierra Region	xvii
CHAPTER 1: THE NATURE OF MOTHER SIERRA	1
Native Soils	2
Precipitation and Water Systems	2
Flora and Plant Communities	3
Exposure	4
Climate	5
Microclimates	7
CHAPTER 2: SO YOU WANT TO GROW FOOD IN THE SIERRA	9
Frosts and Hardiness Zones	10
The Seasons: July and Winter	12
Spring (April 15–June 19)	12
Summer (June 20–August 20)	13
Fall (August 20–October 31)	14
Winter (November 1–April 15)	15
Biodiversity: Keeping in Balance with Nature	16
CHAPTER 3: HOW DOES MY GARDEN GROW?	21
Soil	21
Water	22
Nutrients	23
Organic Composts	24
Manures	27
Fertilizers	29
Cover Crops (Green Manure) and Crop Rotation	31
Mulches	33
CHAPTER 4: ESTABLISHING YOUR GARDEN	35
Site Selection: Locating Your Garden	35
What Do I Want to Grow?	39
Should I Direct-Seed or Transplant?	40
Planting	41
Irrigation	42
How Much Do I Water?	43
Tools for the Garden	47

CHAPTER 5: VEGETABLES, FLOWERS, AND HERBS	53
Selecting Varieties	53
Seed or Transplants?	54
When Do I Plant?	56
How to Grow Vegetables	57
Vegetable Greens	57
Flowering Crops	60
Vegetable Fruit Crops	61
Vegetable Root Crops	64
Seed Saving	69
CHAPTER 6: FRUITS, NUTS, AND BERRIES	73
Fruit Trees	74
Harvesting and Storage	75
How to Grow Fruits	76
Apples	76
Apricots	78
Peaches	78
Pears	80
Plums	80
Planting Fruit Trees	82
Tree Care	84
Specialty Fruits	87
Nuts and Seeds	90
Nuts	90
Seeds	92
Berries	93
How to Grow Berries	94
Wild Plants and Berries	102
CHAPTER 7: PUT ANOTHER LOG ON THE FIRE: EXTENDING YOUR SEASON	107
Site Selection	109
Season Extension Structures	109
Cold Frames	109
Hoop Houses	111
Greenhouses	115
Geodesic Domes	116
CHAPTER 8: MANAGEMENT OF SEASON EXTENSION STRUCTURES	119
Temperature	119
Water and Fertility	120
Establishing Seedbeds	120

Frost Protection	123
Crop Planning	124
CHAPTER 9: PESTS: THE GOOD, THE BAD, AND THE UGLY	129
Wildlife Pests	129
Rodents	130
Skunks	132
Raccoons	132
Birds	133
Deer	133
Bear	134
Insect Pests	135
Indoor Pests	139
Weeds	140
Molds, Fungus, and Diseases	142
CHAPTER 10: A FIRESIDE CHAT AND GOOD NIGHT	143
Appendix A: Sierra Valley Farms Growing Schedules	148
Appendix B: Sierra Valley Farms Companion Planting Schedule	158
Appendix C: Starting Your Mountain Farm	161
REFERENCES	197
ACKNOWLEDGMENTS	199
Index	203

INTRODUCTION

The Sierra Nevada is not the first place you think of when contemplating growing a garden. Who in their right mind would try to grow a garden, let alone try to make a living farming in the Sierra? Well, it's been done, and done successfully. This is my story, and the title *July & Winter* says it all. My uncles and the old-timers in Sierra Valley used this phrase frequently to describe the growing seasons of Sierra Valley, where my farm is located, because it has one of the harshest growing conditions in the West, with only 30 or 50 frost-free days a year. In many years, we have frost every month.

Sierra Valley is located on the northeastern slope of California's majestic Sierra Nevada mountain range. Just under 5,000 feet in elevation, it spans 220 square miles and is the largest alpine valley in the Western Hemisphere. Sierra Valley is known ecologically for its vast plant and animal diversity and geological formations. Sierra Valley overlaps California's Plumas and Sierra counties; my farm is located in Beckwourth, 45 miles north of Truckee and 50 miles east of Reno, Nevada.

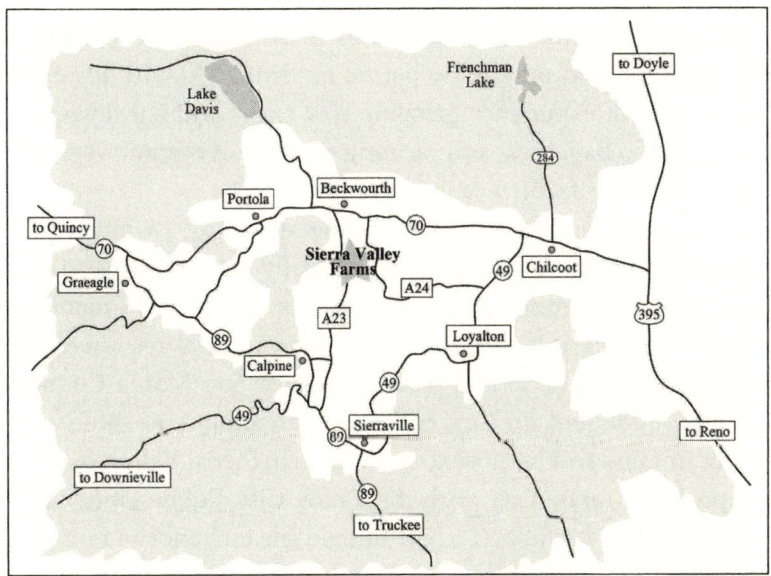

Illustration by Jared Manninen

July & Winter is a sequel to my first book, *Why I Farm: Risking It All for a Life on the Land*, the story of being raised as a third-generation farmer in a hard-working Italian family. That book traced my early years, rebelling and running from farming, and later coming back to save the last family ranch, only to find that farming in the twenty-first century is much different than my childhood memories. In *Why I Farm*, I discussed the challenges of today's farmers, noting that farmers currently make up less than 1 percent of the occupations in the United States. I discussed changes that need to take place in society for a healthier America, and I tried to inspire people to go into farming—it is becoming a lost art, and us farmers aren't getting any younger. Once we're gone, who is going to grow our food?

In my opinion, the only way to feed the world is *one family at a time, one community at a time*—not with factory farms growing one commodity like corn, soybeans, or rice, but by local communities growing a wide diversity of fruits, vegetables, and livestock from local and regional farmers that will sustain our local food sheds. *July & Winter* gives you the tools to grow your own food in the mountain communities of the Sierra Nevada.

Why I Farm inspired me to write this book. I want to help average persons grow food, not lawns, and provide food for their families and their local communities. This book is for those of you living in remote and rural mountain communities who think that you can't grow anything in your garden or can't become a gardener or mountain farmer because of the harsh climates, animals, pine forests, rocks, and acid soils of the Sierra.

I'm here to tell you that you can do it. I've been making a living at it for more than 20 years, and my family before me since 1907. In *July & Winter*, I share my 50 years of experience growing row crops and cut flower orchards in the San Francisco Bay Area, and farming 65 acres of organic vegetables and operating a native plants nursery in Sierra Valley.

When I got to the end of high school, the last thing I wanted to do was farm. But I enjoyed dealing with plants, especially native plants, so I attended the College of San Mateo for an associated arts degree in horticulture, then transferred to Cal Poly in San Luis Obispo for a natural resources degree in parks and recreation. I became a park ranger for San Mateo County Parks, then park superintendent for Coyote Point Recreation Area. But I wanted to live in the mountains and be close to my uncles in Sierra Valley, so I took a job managing parks in Tahoe City with the Tahoe City Public Utility District in 1988. In June 1989, my aunt and uncle offered me a chance to buy the last 65 acres of the family ranch, which I did, and that's how I ended up here.

My first venture was to start a native plant nursery in 1990 called Sierra Valley Wholesale Nursery, where I built three greenhouses, a shade house, and installed a well. I began propagating native plants for highway projects and erosion control jobs, and I started a revegetation and landscaping business to install them, learning how to use native plants in the landscape. In 1995, while I was park superintendent at Truckee-Donner Recreation and Park District, I

Here I am using the same Planet Jr. drop seeder that my dad and grandfather used.

was asked to find a place for a farmer's market in Truckee within our parks, and when I did, the lights went on: Why not make an organic farm out of my ranch, thus the current name Sierra Valley Farms. I knew how to grow flowers, so why not grow vegetables the same way that I was taught by my dad and his father? So I did. I just had to figure out which ones to grow. I'm still trying to figure out what to grow. It's a continuous learning curve.

HOW TO USE THIS BOOK

My goal in writing this book is to give hope to residents living in the Sierra that you can be successful gardeners above 3,000 feet, and, more important, to encourage these gardeners to become farmers who can produce food for our local, rural mountain communities where it is so hard to find good, fresh, organic produce. I give you some options about how to successfully grow your gardens, and I hope to attract new farmers, young and old, and help existing ones produce food for our local communities by way of direct sales at farm stands, U-picks, farmers markets, community gardens, and CSAs (community-supported agriculture).

The farming and gardening practices of this book are intended for areas of the western and eastern slopes of the Sierra 3,000 feet and above. Those of you in the foothills of the Sierra below 3,000 feet have occasional snow and late fall frosts and can pretty much follow the existing *Sunset* guides to gardening, but there are fundamentals in *July & Winter* that can work in your areas as well. My concentrated effort is to help those in the higher elevations who have to deal with unpredictable weather, heavy snow, infertile and rocky soils, and endogenous microclimates.

The book begins with a simple description of the natural history of the Sierra Nevada and its climate, along with its biodiversity of plant communities. Rather than detail the formation of the Sierra Nevada mountain range, I concentrate on the processes that take place at each plant zone and how plants have adapted to the Sierra's harsh conditions.

The book is set up for two types of Sierra gardeners and farmers: those who live between 3,000 and 4,500 feet elevation in the foothill region and those who live between 5,000 and 8,000 feet in the Sierra region, which I designate as the **west** and **east slopes** of the Sierra. Both groups have unique environmental challenges, and I address the specifics to each region. Notes specific to the foothill region are accompanied by an acorn 🌰, and notes specific to the Sierra region over 5,000 feet are identified by a pine cone 🌲. For those of you in the 4,500–5,000 foot range, there is a little overlap depending on your location.

Remember, these are just guidelines, and many of these applications may be effective at any elevation in the Sierra. Experimentation on your own property and situation is the best formula for success. I've survived 20 years of farming food in the Sierra by trial and error and learned more from my failures than my successes.

🌰 Foothill Region

For many people, living in the foothills of the Sierra may mean farming or gardening above the Sacramento Valley floor up to the Auburn, Nevada City, Placerville, and Camino areas at 2,500–2,800 feet. I have a lot of farmer friends in this region, and they don't have the same challenges as we do growing at higher altitudes. According to the *Sunset* gardening guides, below 2,000 feet you can pretty much grow anything you want. I know there are a few fruit tree exceptions, and some of you flatlanders will get on my case for saying this, but hey, it's a breeze compared to farming at 5,000 feet.

My focus is to help gardeners and would-be farmers on the western and eastern slopes of the Sierra who want to grow from 3,000 to 4,500 feet in areas where there is usually frost in October and snow during the winter months. The following areas definitely have four seasons:

- Highway 49 between Downieville and Sattley
- Highway 89 outside of Quincy, including Crescent Mills, Greenville, and Susanville
- Highway 70 corridor east from Beldon, Quincy-Meadow Valley, on to Spring Garden
- Highway 80 from Gold Run, Dutch Flat and Alta east to Emigrant Gap
- Highway 50 around Pollock Pines
- Highway 108 areas around Twain Harte toward the Sonora Pass
- Highway 4 areas below Arnold, east toward Tamarack.

These areas are not inclusive but represent the midrange elevation corridor that extends from north to south along that 3,000–4,500 foot range of the Sierra Nevada.

Those of you in this region fall into the USDA Hardiness Zone 6b-8 and can expect 90–120 frost-free days per year depending on factors like elevation, microclimate, and exposure. If you are pushing the range between 4,000 and 4500 feet you can reduce that to 75–90 frost-free days, and above 5,000 feet you may be reduced to 30–75 frost-free days.

🌲 Sierra Region

The following Sierra regions contain elevations on the west slope mainly ranging from 5,000 to 8,000 feet:

- Highway 70 from Graeagle, Portola, Sierra Valley to Highway 395
- The eastern stub of Highway 49 to Loyalton

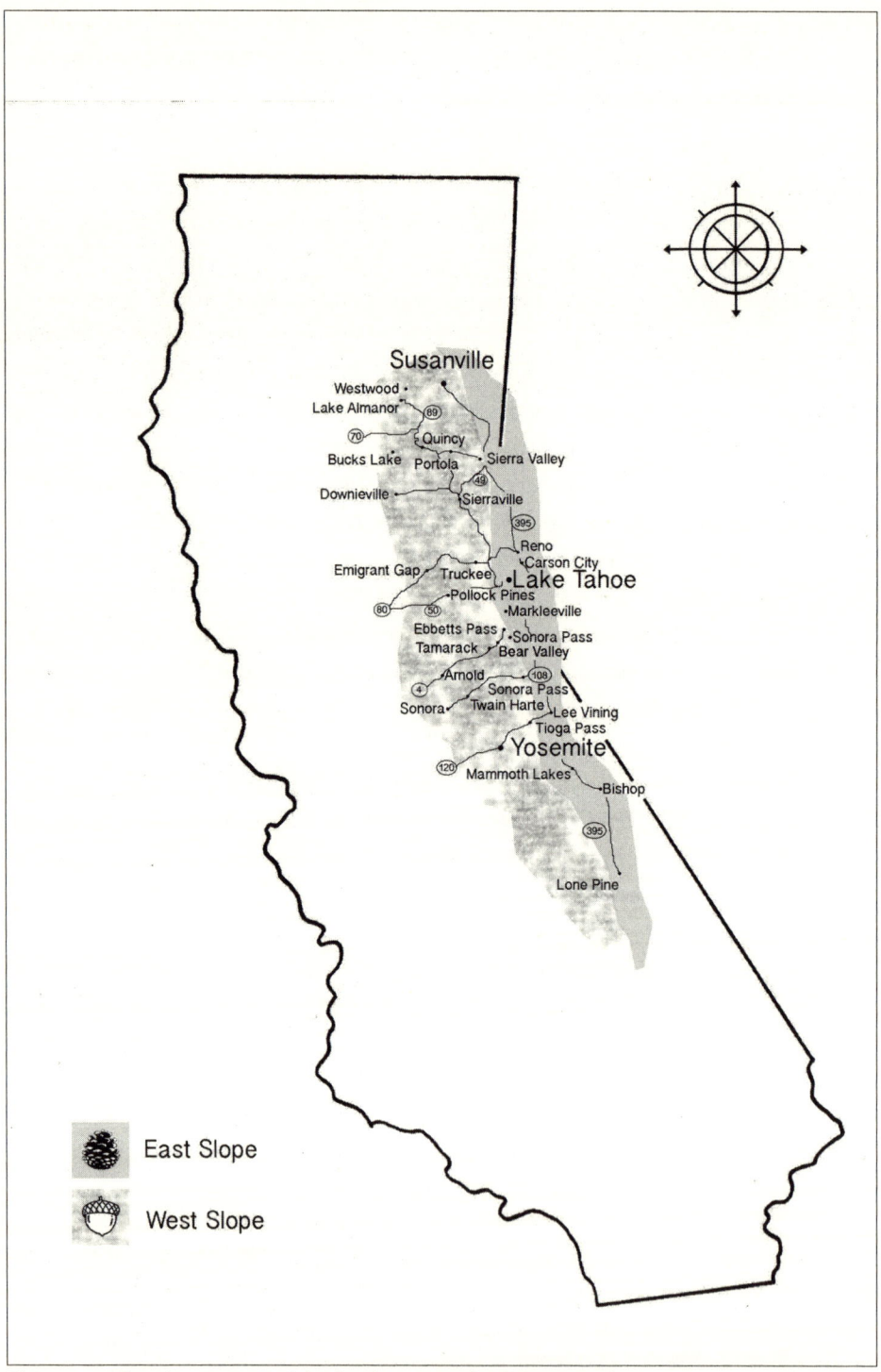

Please note that the contents of this book concern the Sierra Nevada mountain range. The acorn icon represents the 3,000 to 5,000-foot elevations of the west slope, and the pine cone represents the upper elevations (5,000 feet and above) of the west and east slope of the Sierra.

- The northern section of Highway 89 toward Lassen, Bucks Lake, Westwood, the Lake Almanor area, and 89 South including Sierraville, Truckee, and all of Lake Tahoe
- South of Tahoe areas include the western slope areas around Sonora, Ebbetts Pass, Bear Valley, and Yosemite Valley.

On the eastern slope, the Sierra region pretty much follows the areas along Highway 395 above 5,000 in the north with elevations up to about 8,000 feet, including Reno, Sparks, Carson City, and southward through Minden/Gardnerville, Markleeville, on to Bridgeport, and everything around Lee Vining; the east side of Mammoth Lakes and south to above Bishop in the Kings Canyon/Sequoia National Forest region. I've included Tioga Pass and the Mammoth Lakes area because of similar elevation and climatic conditions.

> The key to success is finding the right crop, for the right conditions, for the right time of year.

In the chapters that follow, I help you plan and set up your gardens or farm utilizing the ecosystems surrounding you and build your gardens or farm with biodiversity and organic practices in mind. I focus on soil preparation, land use and designation, soil fertility, crop and variety selection, water and pest management, season extension methods, tools and equipment, and how to harvest your food. There are a number of lists, tables, and illustrations to help you get through the constraints you face when trying to grow gardens and crops in the Sierra. Finally, I conclude with a fireside chat on how to add diversity to your gardens and farm to sustain them for years to come.

The book has a casual and practical approach to farming in the Sierra. I have included my experiences along the way and some fun stories from the past 50 years. I have set up this book for the average gardener as well as the more advanced, and I expect the reader to have some common knowledge of gardening or farming practices. I include some definitions for those of you who are relatively new to the art of gardening and farming. I call gardeners and farmers "artists," because even though gardening principles are based on science, it's the art of the gardener or farmer that in the end sustains the gardens.

I hope that the humor in my successes and failures and memories will inspire you to tackle growing new things in our unpredictable climate. I came to a realization late in life that farming and running a nursery at this elevation is unique, and if it has worked for me all these years, why not try to pass it on to others? There really is no literature out there to help you grow things in the Sierra—yet. I've had to experiment and learn as I go. I can tell you that no two years are the same. You just get a feel for when to do things and hope for the best. I must say that in my years of farming here, as harsh as it may seem, I've never really lost a full crop. Some, yes, but plants are very adaptive to their surroundings if you give them a chance. Too many times we baby them because we're afraid the frost will kill them. The key to success is finding the right crop, for the right conditions, for the right time of year.

Enjoy this book. It was quite an undertaking for me, trying to make sense out of 20-plus years of scrap notes and scribbles and a lot of swearing along the way. Gardening and farming in the Sierra is exciting and fun. The best part is that no two days are alike; it's never boring, it's somewhat frustrating, but every day you face new challenges in trying to work with "Mother Sierra." She has no problem kicking you when you're down, but she rewards you if you can get back on your feet and face her another day. There are many lessons to be learned, and there is no place in the West that can compare to the unpredictable Sierra Nevada. One thing is for sure: You always have July and winter to work with when considering growing food in this crazy place.

CHAPTER 1
THE NATURE OF MOTHER SIERRA

There is no other mountain range in the world like it, situated between the vast Pacific Ocean in the west and the arid Great Basin to the east, a massive uplift of granitic and volcanic rock formations jetting toward the sky. Mother Sierra, the Sierra Nevada mountain range, lies almost entirely in California, extending into neighboring Nevada only along the eastern shore of Lake Tahoe. In Spanish *Sierra Nevada* means "snowy mountain range," an appropriate name because the Sierra is one of the snowiest places in North America. Only the coastal ranges of the Pacific Northwest, from Oregon to Alaska, receive greater snowfall than the Sierra.

More than 400 miles long and 60–80 miles wide, the Sierra forms a massive granite barrier separating central and northern California from more arid lands to the east. The Sierra is the longest continuous mountain range in the West (Whitney, 1979). The Sierra Nevada begins where the Cascades end near Lake Almanor, and they extend to Tehachapi Pass in southern California. The Sierra Crest, which culminates in a single main divide, separates the watersheds of the east and west slopes.

The west slope of the Sierra is a broad, gently sloping ramp, 30–50 miles wide, that rises eastward from the floor of California's Central Valley to about 2,000 feet in the north and 5,000 feet in the south. Above these elevations the west slope is dominated by pine and fir coniferous forest to about 8,000 feet. Despite its relatively gradual incline, the west slope is extremely rugged, intertwined with southwest ridges separating deep river gorges or canyons that were created 20–30 million years ago. Heavy precipitation during the winter created high water flows down these canyons.

On the east side of the range, the Sierra presents an imposing wall, rising steeply from the adjacent lowlands. East slope canyons are shorter and steeper than the west side of the Sierra. Their streams drain into far smaller areas and carry less runoff; the rain shadow effect means less precipitation on the eastern slope. Conifer forests dominate the upper slopes of the east and extend to lower elevations along stream and rivers; they are not as dense as those on the west slope because of the drier conditions. As you descend down the east slope, piñon-juniper woodland and sagebrush scrub grow downslope to the desert valleys at the base of the range (Whitney, 1979). The sub-alpine and alpine regions, which extend above 8,000 feet, reach over 14,000 feet at Mount Whitney—this is known as the High Sierra. Spectacular rock formations, glacial lakes and meadows, harsh winds, and heavy snow accumulation persist year-round at these elevations.

I always wonder what the early explorers thought when they first saw the jagged peaks and large land masses ahead of them. It must have been mind-boggling for Jim Beckwourth, and the Donner Party. First trying to cross this rough terrain in wagons, then trying to settle in this country? And here we are today worrying about trying to grow a garden. Times have changed.

Even so, the majestic geological formations of the Sierra and her climate creates a challenge for gardeners and farmers learning how to grow food in the Sierra. The key is to understand what makes her tick: the dynamics of the Sierra Nevada Mountains. These dynamics include the following:

- The geologic formations (soil production)
- Precipitation and water systems (rivers and streams)
- Flora and plant communities
- Exposure (elevation and slope of the Sierra, east or west)
- Climate (four seasons) and weather patterns
- Microclimates

NATIVE SOILS

The Sierra Nevada is made up of different geological formations. For example, granite is broken down and eroded by changing temperatures (cold, hot, freeze-thaw, wind, sun), creating an acid-base material. Rocks, gravel, sand, silts, and clays are then mixed with duff (a thick layer of pine needles) and leaf/twig litter from evergreen and deciduous trees, shrubs, forbs, and grasses that decompose to create a shallow, acid-base soil. The forest soils of the Sierra in general are very acidic and are low in nutrient value; therefore, when trying to grow food like vegetables it's very important to understand that acid soils are not what vegetables like to grow in—we must amend these soils to be more alkaline. I discuss this in Chapter 3.

PRECIPITATION AND WATER SYSTEMS

Our source of water in the Sierra comes from precipitation in forms of snow, rain, hail, and fog. It all starts at the top and runs downhill. The ideal situation is to saturate the soil to its carrying capacity (when the soil cannot hold any more water), then cover it with a nice snow pack. Warm, gentle spring rains allow the snow to melt slowly and infiltrate water into the subsurface aquifers and the surface runoff to fill our rivers, streams, lakes, and ponds. All these water reserves are the basis for our wells, irrigation ditches, and public water systems. The west slope of the Sierra receives 20–80 inches of precipitation per year, with the majority of the snow depth reaching its peak around mid-March, usually in the 5,000–6,000 foot ranges. The east slope, being more arid, receives only 10–30 inches per year, and mostly in the southern ranges; Mammoth Lakes is the exception, sometimes getting over 500 inches of snowfall in a given year. The Sierra snowpack can range from a minimum of 13 feet (1880–81) to a record 65 feet (1951–52). The greatest annual snowfall ever recorded was at Tamarack (8,000 feet) in Alpine County at 73 feet (1906–07). Wow, I can't imagine farming in a winter like that!

Those of you on the west slope of the Sierra receive the most precipitation because of its gentle slope that extends 30–50 miles wide and includes the Feather River, American River, and Yuba River that empty into the Sacramento River. To the south, the Mokelume, Tuolumne, and Merced rivers all empty into the San Joaquin Valley, and all end up in the San Francisco Bay. The far southern range of the Sierra includes the Kings, Kaweah, and Kern river systems that all terminate in the San Joaquin Valley, and small lakes and playas that never reach the ocean.

Those of us on the east slope have the Truckee River, which is one of the only rivers that flows east instead of west in California. Others include the Carson and Walker rivers that drain on the east slope of the Sierra, a series of tributary streams that deposit into Mono Lake, and to the south the Owens River.

FLORA AND PLANT COMMUNITIES

For those trying to grow food in the Sierra, it's probably most important to learn about the native flora and plant communities of the Sierra Nevada. Understanding how these plants and their plant communities have adapted to their harsh climatic and environmental conditions helps a gardener or farmer adapt crops to their ever-changing surroundings. Native plants in the Sierra that have grown here for thousands of years have adapted to the shallow soils, precipitation allocations, the climatic elements of wind, rain, cold, heat, snow and ice, exposure to the sun's east-west-north-south–facing slopes and the major seasons: July and winter.

We begin with the Sierra foothills, in which the elevation below 3,000 feet includes oak woodlands, digger pines, big-leaf maples, poison oak, gray manzanita, and buckbrush. The **lower montane**, the forest-belt, mixed coniferous forest elevation between 3,000 and 6,500 feet on the west slope consists of ponderosa pines, black oaks, big-leaf maples, aspens, cottonwoods, Douglas and white fir, Jeffrey pine, manzanita, buckbrush, ash, and dogwood, to name a few. The east slope consists of western juniper, piñon pine, Jeffrey pine, cottonwood, scrub oaks, sagebrush, mountain mahogany, bitterbrush, rabbit brush, serviceberry, and desert peach, to name a few.

In the **upper montane**, 6,500–8,000 feet, the soils are shallower and the environments are more extreme. The wind shear on the mountain tops and ridges increases in velocity as the winds are compressed upward, temperatures drop, and only the strong survive. These native plant species survive by growing shorter, living in crevices behind rock outcroppings, and having shorter needles, a thicker, more narrow branching, and a heavy, waxy cuticle coating on their leaves to retain moisture and prevent frost burn. These species include red fir, mountain hemlock, lodge pole, limber (east), foxtail, and white bark pines, western juniper (east below 7,000 feet), mountain heather, huckleberry oak, whitethorn, snowbush, and pine mat manzanita.

Everything above 8,000 feet is considered the alpine region, and anyone considering gardening above 8,000 feet would be doing so in an artificial condition, like a growing dome (see Chapter 7), or if not, should have their head examined. It's way too extreme to sustain food at that elevation, in my opinion.

The more you understand how native plants have adapted to Mother Sierra's native garden, the better you will be able to design your garden or small farm in the Sierra. The native flora of the Sierra are grouped in plant communities; each plant community is grouped together because of its adaptability to Mother Sierra's elevation, exposure, slope, water availability, soil moisture, soil type, and adaptability to the climate (including snow, wind, rain, heat, cold, shade, and microclimates).

We recognize plant communities in the Sierra mainly by elevation and slope. I have studied native plants since my days at Cal Poly and into my Parks and Recreation careers at Truckee and Tahoe City, taking my trade from the learning stages into practice, owning my own native plant nursery and landscape and erosion control business at Lake Tahoe in the 1990s. I can tell you from experience that native plants are very good indicators of elevation and soil moisture. Certain trees, shrubs, forbs, and grass species will grow only at certain elevations and under certain soil moisture conditions. Not only that, but incorporating them into your gardens will entice pollinators, beneficial insects, birds, native bees, and beneficial predators to your gardens.

EXPOSURE

Exposure is a combination of climatic conditions, elevation, and the position of the slope of the Sierra Nevada Mountains. The volatile climate of the Sierra is affected by the position of its slopes. The rise in elevation starts with the foothills and rises sharply to more than 14,000 feet at its peak.

Because the Sierra is so abrupt with extreme uplifts, everything that's on the top of the mountain will eventually end up on the bottom. That's just gravity, folks. So our alpine 100-mile-per-hour winds, sleet, heavy snows, ice, and freeze-thaw begin to break down these granitic forms, and as rocks break into sediment and tree limbs break and fall over and break down into humus to create soils, it all washes down the mountains and native plants grow from it as they adapt to the climate's conditions.

When you look at your own property, the first thing to do is consider where you're located in the Sierra. This may seem obvious, but I don't mean street, address, city, and state—I mean elevation, situation of slope (east or west slope, because the Sierra Nevada Mountains run north–south, so your weather exposure and elevation is either on the west slope looking toward the Pacific Ocean, or it's on the east slope looking out to Nevada).

> Exposure to slope is a major consideration when deciding what to plant on your property.

Exposure to slope is a major consideration when deciding what to plant on your property. South-facing slopes receive more hours of sunlight and more intense radiation than those facing north. As a result, they tend to be warmer and drier than northern exposures. Northern slopes tend to face away from the sun in the Northern Hemisphere; these slopes are colder, thus holding the early winter snows, and the last to melt off in the spring. East- and west-facing slopes exhibit less microclimate differences than the north–south phenomenon. Both receive comparable amounts of sunlight during the day. Western exposures tend to be somewhat warmer, while eastern slopes are drier and steeper because of the Sierra geological formations and may lie in the shadows of the western slopes because of the sun setting in the West. Eastern slopes are generally cooler than western slopes, especially during the winter months when the sun is low in the sky.

CLIMATE

One of the most important things about working with Mother Sierra is understanding what makes her tick. As neurotic as she can be, there is a method to her madness. As they say in the mountains, "If you don't like the weather, just wait five minutes and it will change."

The Sierra Region has two distinct climates in reality: the west slope, and the east slope. Both are very different in the amount of snow fall, precipitation, and exposures they receive. Remember that the upper-west slope areas get more snow and precipitation than the eastern slopes, but the eastern slopes are colder in the winter and hotter in the summer, less forested, and more arid.

The east slope tends to have high pH water and alkaline-sandy soils, whereas the west slope facing the Pacific Ocean receives more snow and precipitation, has heavy forest cover attributed to acid water, and has rocky, shallow, acid soils. Even though the elevations may be the same, the growing conditions are totally different from east to west.

Climate and the seasons change very quickly and often in the Sierra. If you want to garden or farm like I do, you can't be a Type A personality, unable to adapt and go with the flow. The only way I've been able to farm successfully at 5,000 foot elevation is as a Type B personality, where I can change course hourly, daily, and weekly, adjusting to Mother Sierra and her crazy climate changes. My actions on the farm when planting, harvesting, and selecting species depend on what's happening with the weather. Right or wrong, my decisions are 90 percent based on what forecasters say she's going to do this week. I have kept weather logs for 25 years here on the farm, and I can tell you that no two years are the same, there are no patterns, and weather can change at a drop of the hat at any time of the year. But there are some guidelines to follow.

> Microclimates within a garden or farm give an opportunity to the gardener or farmer to grow specialty crops that would normally not grow there.

One element that really affects temperatures is elevation. This phenomenon of mountains is amazing to me. Mountains, because of their heights, are wetter, windier, cloudier, and cooler than lowlands. Large mountain lakes and valleys like Lake Tahoe and Sierra Valley can create their own weather, which can change rapidly. Mountains work like this: As moist winds are forced upslope they become condensed and cooled to the point that the air gets thinner and water vapor creates clouds, which become raindrops or freeze into snowflakes. In general, temperature drops 3.5 degrees F (temperatures throughout the book are given in Fahrenheit) for each 1,000 feet of rise in elevation, and precipitation increases 2–4 inches for each 300-foot rise, thus reaching its maximum between 5,000 and 6,000 feet in elevation. That's why the west slope of the Sierra at this elevation receives the most moisture around Blue Canyon and Kingvale. After that elevation, precipitation declines as the mountain air becomes drier and is influenced by the arid (desert) east slope of the Sierra.

The combination of heavy snowfall and below-freezing temperatures are characteristic of Mother Sierra's winters, and 95 percent of her total annual precipitation occurs between October and May, with January the coldest and wettest month on average. The heaviest snowfall usually occurs because of the Hawaiian storms, and most storms average about 6 days. The most extreme winter temperatures occur around the Truckee/Boca area, where temperatures can reach well below –20 degrees, with the majority of the snow loads occuring up around the Donner Summit and Kirkwood areas.

Microclimates

The last point I cover in this chapter is probably most important for understanding how to grow food in Mother Sierra's gardens, and that is utilizing the microclimates that occur everywhere in the Sierra and at all elevations. Microclimates are small pockets of specialized environments caused by a combination of temperature, moisture, humidity, wind, sun, shade, vegetation type, and exposure. Microclimates within a garden or farm give an opportunity to the gardener or farmer to grow specialty crops that would normally not grow there because of normal cold or warm conditions relative to the specific elevation. These pockets of specialized environments help protect the specialty crops.

Some microclimate basics:
- Heat rises, cool air falls, thus upper slopes are warmer, valleys and meadows are cooler.
- Trees create windbreaks, shade, humidity and warming effects.
- The sun comes up in the west and sets in the east, and is lower in the sky in the fall and winter.
- Evergreen trees keep their leaves all year and provide windbreaks and shade in summer and winter, but hold snow longer in spring.
- Deciduous trees give windbreaks and shade in summer, lose their leaves in fall, and allow sunlight through the branches in winter, thus allowing for faster snow melt in spring.
- Ponds, meadows, and bogs contain more soil moisture, attract more insects, and have higher humidity in spring and summer, but are colder at night during winter and are windy.

During the summer, rain causes high humidity. A hot summer day will cause a sudden thundershower to drop 2 inches of hail in 1 hour, then clear up to a starry night and a killing frost the next morning. A drastic change in summer temperatures can cause unsettling winds that create a microburst, where winds become devastating "dust devils" that rise the air, then rapidly thrust downward to demolish small buildings and structures at a moment's notice. These are just a few examples of microclimates. Case in point: In summer 2014, we had a really warm calm day until about 3:00 p.m., when all of a sudden a swirling wind began and a dust devil across the road from our farmers market area gained speed, came across the road and into our courtyard, sucked up an umbrella from one of our tables, twirling it into the air about 50 feet, only to turn into a downward burst that sent the umbrella like a harpoon through my 50-foot greenhouse, exposing a gaping hole that the vicious wind proceeded to peel the plastic film off like a glove, leaving only the skeleton of the greenhouse. Mind you, in front of my eyes this all occurred in about 20 seconds!

The point is to understand that microclimates, good and bad, are going to happen and happen in different ways at different times of the year. It is very hard to plan for them, but understanding the nature of Mother Sierra and what makes her tick will help you succeed in growing food in the Sierra.

CHAPTER 2
SO YOU WANT TO GROW FOOD IN THE SIERRA

So, how do you fit in the Sierra? This chapter is all about evaluating your property's location and how you fit into the Sierra landscape. You will look at your elevation and exposure to slope, identify general native plant communities, and which planting zones you are situated in. You'll also consider the seasons and biodiversity.

Begin by identifying your side of the Sierra, west slope or east slope, and your elevation. Identify the native and non-native plants in your region as much as possible. (A good way to do that is contact your local chapter of the California Native Plant Society or Master Gardener Program and become involved with your local group.) Look around you and identify the plant community you're situated within. Identify dominant plant species in your plant community like ponderosa, or Jeffrey pines, white fir or red fir, broad-leaf trees like oaks or maples, and aspens or cottonwoods.

Why is this important? Because identifying these plant communities and plant species identifies your elevation. Native plant communities are a make-up of several plant species that have similar growing conditions—for example, soil type, moisture condition, slope exposure, and sun exposure. Knowing what forest plants are around your property will help in placement of your gardens and will allow you to select plants that are compatible to your surroundings. It will also help you to research what wildlife and insect species that are both good and bad for your gardens live in the plant communities. Plant communities give you a feel of where you fit in the Sierra. In addition, once you've identified a plant community or looked up an individual plant species online, you can find out its height, water requirements, botanical and biological make-up, and more, so that you can add them into your landscape for wind breaks, shade, wildlife attractions, beneficial insects, and accents. Next, identify your microclimates—for example, meadows, ponds, slope exposures (east, west, south, north), predominant wind directions, and shade and sun times of the day. Take a look at the wildlife that inhabit your farm or gardens, what they eat, where they live, and what types they are (birds, rodents, amphibians, large animals, pests or prey, the good, the bad, and the ugly).

Now that you've established where you fit into the region you want to farm or garden in, and have an idea of how volatile Mother Sierra can be, you must set up your garden in a way that can grow food with the elements Mother Nature has given us. We know that outdoors we only have a small window of frost-free days to work with, and from then on during the cooler times of the year we are going to have to amend our practices in order to grow food year-round. (In Chapter 7, "Put Another Log on the Fire: Extending Your Seasons," I discuss options for gardens to augment Mother Sierra's continual weather changes.)

FROSTS AND HARDINESS ZONES

When trying to grow food in the Sierra it is best to start with some common knowledge and generalities about growing food. The first thing to do is to establish which hardiness zone you are located in. For the Sierra Nevada mountain range, the USDA Hardiness Zone falls somewhere between zone 5a-6a 🌲 up to zone 6b-8 🌰. You can go on the USDA Hardiness Zone website (www.planthardiness.ars.usda.gov) to get the exact hardiness zone for your area. These zones give you a ballpark rating when you are looking up specific ornamental and fruiting shrubs and tree varieties for your region.

For those of you starting your gardens or small farms at or above 5,000 feet region (like me), we are considered in the USDA Hardiness Zones 6a-6b (old classification of Zone 1-3) that have only 50–75 frost-free days per year; some of us who live in high mountain valleys or around Truckee/Tahoe may even be in the 5a-b range and have frost every month of the year. I've seen it many times in Sierra Valley.

I've tabulated some averages based on statistics from the Soil Conservation Service soil report (1930–64) of Sierra Valley; Young's *What Grows Here in Plumas, Sierra, and Lassen Counties* (1973–82); and my 20-year logs here on the farm in Beckwourth (1989–2012). To see frost-free day averages tabulated over a period of years by national weather stations in the surrounding areas see Table 2.1: Average Frost-Free Days per Elevation.

In the Sierra many of the seasons overlap so much that it's hard to tell when one ends and the next one starts. As I'm writing this book, it's December 23, 2014, two days after the solstice, and it's 40 degrees in the morning with a high of 63 degrees and bright sunshine. Last year on this date, the temperature reached −18 with a high of −2. You'll have better odds at a crap table in Reno than predicting common years in the Sierra.

The USDA Hardiness Zones do very little for vegetable production, because your main concern here is **frost-free days**. What makes frost-free days challenging in the Sierra is that in most cases reports are given for consecutive frost-free days, and that is not the case in the Sierra. Very rarely do we have a whole growing season with a lot of consecutive days without frost. We tend to have split seasons. I can say from experience in Sierra Valley, we will get warm late May through second week of June, say 17–18 days of great weather to get you excited to put those annuals out and then *bam!* . . . a couple of 25 degree days, then we're good to go without frost usually until maybe late July or August 1, then again without warning *bam!* . . . a 30-degree day or two, then warm weather through Labor Day.

Table 2. 1 Average Frost-Free Days per Elevation	
Elevation	Frost-Free Days
3,000–4,000 feet (west slope)	90–120 days
4,000–5,000 feet (west slope)	60–90 days
4,000–5,000 feet (east slope)	40–95 days
5,000–above (east/west slope)	50 days or less
7,000-above	frost every month
Specific Averages	
Quincy (3,400 feet)	86
Susanville (4,148 feet)	104
Chester (4,525 feet)	60
Mineral (4,875 feet)	45
Sierraville (4,975 feet)	66
Beckwourth (4,950 feet)	41
Shasta Lake (5,850 feet)	50
Tahoe City (6,230 feet)	71
Note: Table based on statistics from the Soil Conservation Service soil report (1930-1964) of Sierra Valley; Young's *What Grows Here in Plumas, Sierra, and Lassen Counties* (1973-1982); and my 20-year logs here on the farm in Beckwourth (1989-2012).	

That's what kills you in the Sierra: unpredictability. Based on elevation and exposure, the higher in elevation you go and the more you're exposed, the shorter your growing season. You are growing in a cool and semi-arid to subhumid region. The winters are cold and the summers are dry with large temperature fluctuations during the day (as much as 40–60-degree differences between sunrise and midday), cool mornings, and hot afternoons.

According to my ledgers over the past 25 years these are some ranges that I follow:
- In the foothill or Sierra region I base a killing frost at 28 degrees instead of 32 degrees because I grow cool-season crops outdoors without row covers (brassicas, spinach, kale, lettuces, and root crops), and they tend to be hardier than warm-season crops (squashes, corn, tomatoes, peppers).

- My last killing frost is based on June 20 (first day of summer), and my first killing frost is around August 10, which gives me 50 frost-free days of security. On each end I cheat two weeks earlier or later depending on the spring or fall conditions predicted that year. So far this theory hasn't proven me wrong. I haven't lost a full crop in 20 years! For those of you downslope you can increase your growing season by extending your last frost-free day to May 31 and your first frost being at the end of September, in general. I follow weekly National Oceanic and Atmospheric Administration (NOAA) reports religiously, as right or wrong as they may be, it's a guide and better to plan accordingly.

THE SEASONS: JULY AND WINTER

When gardening in the Sierra, July may be your only frost-free month. Spring and fall each carry a hint of winter that can spoil the harvest for many of your early and late summer crops. Understanding the seasons of the Sierra and what each brings is a must, because Mother Sierra will surprise you, that much I can guarantee. Next I describe how I interpret July and winter, or the four seasons of the Sierra for those of us at 5,000 feet and above. For those of you in the 3,000–4,999 foot range you can safely add two weeks on either side.

Spring (April 15–June 19)

Spring is a very unpredictable time in the Sierra above 5,000 feet. You can pretty much expect March–April to be a lion, with some snowstorms, 1–2 feet accumulations of snow, and go out like a lamb in late May. Many times April is cold, still in the teens and 20s, warming up to mid-50s in the afternoon. April brings some rain, and a lot of wind that starts drying out the lower mountain valleys. In some years, April and May can be deceivingly warm, like we are going to have an early summer, getting into the 70s and even 80s for weeks at a time, and everyone gets excited and puts in their annuals and tomato starts and Mother Sierra says, "Gotcha!" and a cold front quickly moves in over Memorial Day and zaps all your plants. That's why in spring I usually drag my feet on putting out any transplants until after June 20.

In April, May, and June the days start getting longer, the days warm gently, but the nights are still cold with heavy morning frost. Mother Sierra doesn't quite want to let go of Old Man Winter. It's "mush" time for me—everything is melting and running off and there's not much I can do with my soils outside other than prepare my beds.

There is typically an intermittent storm right when things are going to dry out, and then comes the wind . . . oh the wind! Mother Sierra is trying to get things dried out and just when she does, she'll give you a warm spell in May to tease you that summer is around the corner and while you're sleeping at night she'll whisper in your ear . . . "It's okay to plant your flower beds and tomatoes now." So you get up and run to your local nursery and put in all your petunias and heirloom tomato plants and . . . *zap!* The next two weeks in June it's 25 degrees and it kills everything that you've planted . . . as Mother Sierra laughs . . . *heehee!*

🌰 I spend most of my spring seeding cold-season crops (spinach, lettuces, radishes, carrots, arugula, kale, collards, Swiss chard, herbs, and Asian vegetables) in my hoop houses (low tunnels or high tunnels); germinating my starts in the greenhouse; pruning down my perennial beds; planting bulbs, spring garlic, onion sets, horseradish tubers; and preparing my raised beds and fields for summer production.

Summer (June 20–August 20)

Frosts are regular in June and then again in August, especially above 6,000 feet. July is typically your only safe frost-free month. Here in Sierra Valley, because of our vast open space and surrounding mountains we have frost most years every month of the year. Here our temperature lows are many times as cold as or colder than Truckee, so it's nothing new for me to expect frost at any time. These are my adjusted dates for summer. If you can go this long without a frost you've had a good season. 🌰 If you are between 3,000 and 5,000 feet, depending on many factors, you can extend your season two weeks earlier in June and sometimes up to a month into the middle of September.

As the Sierra summer temperatures increase, our morning temperatures still stay in the 40s and 50s depending on our elevation, with our highs usually in the high 70s and high 80s. July is usually our hottest month, reaching the 90s and a token 100 once or twice a year. Summer thunderstorms are frequent, usually in the mid-afternoons, and help cool down temperatures. Thunderstorms usually develop within 20–30 minutes and are isolated showers that can dump a lot of precipitation in forms of rain, hail, and snow, sometimes all at once, and are associated with lightning. These lightning storms are great for your gardens because the electrons in the atmosphere produce a nitrogen ion (nitrous oxide) that acts like a fertilizer for plants. I swear I can almost watch my lettuces double in size overnight after one of these lightning storms. Summer winds in the Sierra are very common, daily southwesterly winds start about 11:00 AM in the morning and continue well after 5:00 PM.

🌰 Many times those of us on the east slope, with the influence of the Great Basin, find our morning temperatures in July in the mid-30s and by 3:00 PM pushing into the 90s with a strong wind. This can create a nightmare for your row crops or season extension structures because they must be able to withstand strong winds and have plenty of ventilation. You'll find yourself shutting down your vents, or closing them at night in case of frost, and opening them wide open in the late morning to cool them off. Remember, above 5,000 feet, you are a mile closer to the sun, the heat intensity is much more dramatic and extreme, and within hours you can frost your plants in the morning or fry them in the afternoon.

About 10 years ago I decided to save some money by not using a greenhouse and I shut off the electric fans during the summer. When I went back, in early September, it had gotten so hot inside that it melted all my PVC irrigation lines; they looked like wet noodles. It must have gotten up to or over 180 degrees! I didn't do that ever again. On the other extreme, in 2012 I had a beautiful crop of tomatoes, peppers, and cucumbers ready to harvest in late July. We had been having record temperatures of over 95 degrees in the afternoon, so I was leaving the sides of the hoop house open to allow them to vent early in the morning, and without warning, one morning a cold snap of 24 degrees whipped in for just a few hours and froze my crop! That instance taught me a lesson to never let my guard down and to close up the greenhouse every night. That's Mother Sierra, kicking me in the ass for being lazy! The key is, Pay attention daily.

As summer comes to a close in September, the sun sets low in the sky, sunsets are great, the winds are calm, and there is just a touch of Jack Frost nipping at your nose with daily temperatures in the high 60s and 70s. That's when my cool season crops do best outside (without row covers). My lettuces, kales, carrots, radishes, spinach, mizunas, and arugula flourish in abundance, and I know winter is just around the corner. It's time to start seeding my hoop houses and greenhouses.

Fall (August 20–October 31)

The winds begin to die down, the sun lowers in the sky, and oh, those fall colors! It reminds me of deer season when I was a boy. I would come up from the Bay Area after hay season in September to hunt deer with my uncles here on the ranch. You could smell the morning dew on the cut stubble of the grain and alfalfa fields and the pollen of the rabbit brush and sagebrush in the afternoons as we combed the mountainsides for that trophy buck. The uncles were definitely mountain men, deadly with an open sight rifle, and deer season was an important part of their livelihoods. As ranchers they harvested 5 or 6 deer every year to put in the freezer that fed Grandma and the families of five brothers. To this day I love fall because of those memories and the fall colors. It's my favorite time of year.

🌰 Because of the imminent danger of frost, I've shortened my summer dates to about August 20; if you haven't had frost at the upper elevations, you will any day now. I just get a fall feeling in late August as the days start getting shorter that it's time to plant my winter crops in the hoop houses and my last cold-season crop outdoors. If you have tomatoes and peppers, squash or other summer vegetables still growing in the hoop houses, you will need to start protecting them with some frost protection fabric if your outdoor temperatures are getting close to freezing. For winter-spring carrots I like to get them seeded in the hoop house by mid- to late August for harvest in mid-December through March and April. I seed all my cold-season spinach, kales, lettuces, mizuna, collards, radishes, and chard around Labor Day in my hoop houses and continue succession plantings until Thanksgiving.

If you are not planning on a winter garden, this is the time to seed a cover crop in your hoop house for early spring cultivation, or amend your soils, add fresh manures, straw, and composts then turn them under to compost over winter, so your soils will be ready for early spring (February–March) plantings. Many times if you are growing late-summer vegetables like tomatoes and pepper they will persist until almost Thanksgiving if protected with fabric cloth.

Winter (November 1–April 15)

Ahhh . . . Mother Sierra's favorite time of year! It is possible to grow through winter in the Sierra, but it is a lot of work for what you get out of it. (Please refer to Chapter 8 for details on growing food for the winter months.) If you plant your 30–45 day cold crops, such as lettuces, radishes, spinach, kale, chard, collards, mizuna, and Asian vegetables, in September, and successively plant them again in October and November, you can harvest comfortably until mid- to late December and into January.

In general, November is a good time to compost and add amendments to your soils, turn them under, and then add 3–4 inches of straw or pine needles over the top to insulate them through the winter so they will be ready for spring. Winter is a good time for cleaning collected fall seeds, going through your seed catalogs and making your wish lists of seeds and plants, and then ordering them. It's when I fix equipment that I broke, perform equipment maintenance, and just unwind and enjoy the holidays. Not much grows between December and the end of January, but I may keep a few rows of kale, spinach, mache, and chard going under row covers in my hoop houses for a few of my restaurant accounts. But, overall, it's regrouping time for me, and a time to reflect on my successes and failures of the year and get excited about my crop planning for the upcoming year.

One thing that I recommend is to establish a weather station at your house or farm. It can be as simple as installing a rain gauge and an inside/outside digital thermometer to keep records of seasonal highs and lows (temperatures, snow, rain, hail, thunderstorms, high winds, and so on) and keep logs on special events that Mother Sierra throws at you.

It's a fun thing for the kids to do during the winter, and it keeps them interested in gardening and paying attention to Mother Sierra. I've tabulated the past 25-year average temperatures here on the farm, and its fun to look back on them (see Tables 2.2 and 2.3).

Table 2.2 Average Temperature and Precipitation per Month Beckwourth, California			
Month	Max T	Min T	Precipitation
January	40	12	2.5
February	43	16	2.8
March	50	21	1.5
April	58	26	1.2
May	65	31	1.0
June	71	35	0.2
July	87	40	0.1
August	88	35	0.1
September	77	33	0.3
October	70	28	1.4
November	51	19	2.0
December	41	15	4.0
Total			17.1
T = temperature; precipitation measured in inches			

BIODIVERSITY: KEEPING IN BALANCE WITH NATURE

When I talk about keeping in balance with nature, I mean growing food with whatever weather conditions Mother Nature gives you and understanding how to use the tools you are given. Over the past 5 years I've really taken time to reflect on my 50 years spent farming and gardening, and I believe that gardens and farms are places that are in balance with nature. They are places

Table 2.3 Sierra Valley Farms Annual Weather Statistics	
Lowest winter temperature (1990)	−28 degrees
Average winter low temperature	−8 degrees
Highest summer temperature (2009)	104 degrees
Heaviest snowfall season (2011)	18 feet
Shortest season: (2012; 2011; 2010)	Frost every month
Average precipitation per year	17 inches
Average frost-free days per season	41 days

that are sustained with minimal human inputs and influences that persist on the virtue of their coexistence with microorganisms, plants, and animal life—that is, biodiversity. It's as simple as that, but very complex when you try to put one into existence from scratch, or modify one that has been neglected or abandoned.

Nature surrounds your farm or gardens with native plants that provide microclimates, food, and shelter for a variety of animals and insects, good and bad. The key is to create a series of checks and balances on your property. This includes considering the predator–prey relationship—the good, the bad, and the ugly. Different perennials, annuals, shrubs, and trees that flower throughout the season attract beneficial insects and predators; remove those habitats that attract pests to your gardens, especially rodents, and create habitats that will keep predatory animals (such as owls, hawks, snakes, amphibians, foxes, bobcats, coyotes, and bats) around to take care of the voles, mice, gophers, ground squirrels, chipmunks, and other critters. You have to decide what you want to live with. In many cases you may not want the snakes, foxes, and coyotes around because of your cats, dogs, or livestock. Then, seclusion of your garden may be another option.

Understanding and assessing biodiversity resources and organic practices in the garden is critical. Organic operations generally rely on biodiversity to increase health, vigor, and resilience of the soil and crops. Mother Sierra's native flora gardens are sanctuaries that host beneficial insects, amphibians, birds, and mammals, along with the help of soil-borne microorganisms to assist our gardens in growing food as productively and efficiently as possible. Creating and maintaining wild habitat can be enduring and cost effective; an organic garden is measured not only in yields, but also by the quality and biodiversity of all life.

Gardens and farms are ecologically diverse locales, and this contributes to the health of our environment. How many different species can you identify in this photo? In your garden?

Some ways to increase biodiversity within and around your garden or farm include the following:
- Protect and enhance endangered and threatened plant and animal species and their habitats.
- Conserve and restore native plant communities within your property.
- Create and restore wildlife buffer corridors with diverse plant species of trees, shrubs, flowering perennials, and native grasses.
- Create and protect bodies of water (waterways, meadows, seeps, and ponds) on your property for wildlife and beneficial insects.
- Eradicate non-native weeds and invasive species within your property.
- Prevent erosion of stream banks and gullies and washes using native seed mixes and soil erosion–control fabrics and rock to prevent sedimentation and pollutants into rivers and streams.
- Avoid conversion of sensitive habitats to agricultural production.
- Place unused blocks of native lands into agricultural conservation easements and get paid to protect the Sierra.

These are just a few ways to create, protect, and enhance biodiversity of your garden or farm. For more detailed processes on how to do this refer to these resources:

ATTRA (National Center for Appropriate Technology, NCAT; www.attra.ncat.org)

National Invasive Species Information Center (NISIC) (https://www.invasivespeciesinfo.gov)

National Wildlife Federation (www.nwf.org)

NRCS-USDA Natural Resources Conservation Service (www.nrcs.usda.gov)

Wild Farm Alliance (www.wildfarmalliance.org)

The dynamics of an organic farm or garden lie in creating a world that lives in harmony with itself and its surroundings for years to come, a sustainable plant community with minimal inputs from humans. Remember that your gardens and farms are living things; you need to monitor your development by continually replenishing your nutrition program with composts and manures and diversifying your surroundings, continuing a permaculture of flowering annuals, perennials, and native shrubs and trees.

In the end, your farm or gardens should have a peaceful, rewarding feeling. A place at peace that you can sit out in the evening with a glass of wine and listen to the birds chirp and watch rows of flowering perennials host numbers of bees and butterflies. You can hear the whistling wind and see the sunset's glow on the tree leaves, glimmering on a warm, summer evening. As you look at the fruits of your labor, that beautiful row of spring rainbow chard or clusters of red tomatoes and rows of corn in the summer, or those bright orange pumpkins amongst the fall foliage backdrop, you'll enjoy your efforts, knowing that winter is not far off.

CHAPTER 3
HOW DOES MY GARDEN GROW?

Now that you understand how your location works with your specific Sierra elements, let's consider the major components of soil, water, compost, cover cropping, and the use of fertilizers.

SOIL

The sustainability of a garden or farm begins and ends with the soil. What happens in the underworld (soil) is the life and blood of all agriculture. Soil is defined by the Natural Resource Conservation Service (NCRS) as "the unconsolidated mineral or organic material on the immediate surface of the earth that serves as a natural medium for growth in land plants. Soil's upper limit is the boundary between soil and air, shallow water, live plants or plant material that have not begun to decompose."

In farmer terms, soil structure is made up of sand, silt, and clays along with organic matter that creates loams (blends) that are our growing media for fruit and vegetable production. The texture of each feels very distinct as you crumble soil between your fingers. Sand is the largest particle; it is coarse, loose, nonbinding, and gives drainage and aeration to the soil structure. Next are silts, which are finer, smaller particles that have a slight silkiness to them with a little grit, and when squeezed will bind together somewhat in a ball; they are essential for water- and nutrient-holding capabilities. Clays are the finest particles, very sticky—like the old Play-Doh we used to play with as kids—and will make a flat ribbon when squeezed between your forefinger and thumb. Clays have the highest water- and nutrient-holding capabilities but bind very tightly, not allowing air between the particles.

For your gardens, the optimum condition is a soil that has equal parts of sand-silt-clay that contain 3–5 percent organic matter. These soils have good structure with ample nutrient and water-holding capacity. Sandy or silt loams are preferred for most vegetable production areas. Clay loams tend to become wet boggy areas in the winter and are slow to dry out in the spring. Improper irrigation practices during production can cause fungal and bacterial problems in vegetables and winter molds because of lack of aeration.

So how do you know what type of soil you have? The easiest way is to make a profile or texture of your soil. Take a quart Mason jar and fill it two-thirds full of water, add 1 teaspoon dish washing soap, then add your soil until full, fasten the lid and shake materials vigorously on and off for 10–15 minutes, then let settle for 2–3 days. The sand particles will fall to the bottom, the next layer will be your silts, and the upper layer will be your clays. Each layer will have a distinct color separating the three types. The sand will fall out in the first 5 minutes, then your silts take about 2 hours, and your clays will take anywhere from 1 to 2 days. You can mark each layer with a Sharpie pen, and then you can estimate your percentage of sand-silt-clay loam. The highest percentage dictates your loam, for example, silty-clay loam, or sandy-silt loam.

From there you'll need to test your organic matter, pH, and nutrient levels of your soil. For your gardens, a pH range of 6.0 to about 7.5 is ideal, with 7.0 being neutral. What do these numbers mean? Well, here is the scale: 4.5–6.0 are very acidic soils, these are your forest soils where not much grows other than blueberries and pine trees. Ever notice that not much grows under pine trees? That's why . . . it's too acidic. The other extreme is a pH of 7.6–9.0. These are the very alkaline soils you see out in Fallon, Nevada, or down in southern California—very sodic or salt-flat areas where nothing will grow because they are very heavy clay-hardpan areas. The Great Basin–type soils on the east side of the Sierra lack rainfall that would leach out a lot of these alkaloidal clays to make them more acidic, so when these soils reach a pH of 8.0 or higher some of the iron, phosphorous, zinc, and manganese become unavailable to the plants. Iron deficiencies, or a symptom called *chlorosis*, then become a problem in annuals and woody plants, and the only way to correct these soils is by opening them up with gypsum and large amounts of composts, organic matter, and composted manures from acid soil farms.

Soil tests are the only way to test for pH levels. A soil test will also test for many other items of interest like organic matter percentage, macro- and micronutrients (see the following nutrients section for details), salinity (salts), electrical conductivity and cation exchange (CEC), which is the ability of the soil to hold nutrients and make them available to plants.

It is best that every 3 years you take soil samples to see what you have to work with and how your soils are developing. I use Peaceful Valley Farm Supply in Nevada City for soil tests, or you can ask your local nursery, Ag Commissioners office, or University of California Extension Office; the test costs about $30, and it takes about two weeks for the results. Just bring them about a pint or so of soil in a brown paper sack and they will send it in for you.

If your pH is below 6.0 you must consider adding alkaline organic matter of leaf composts, straw, manures, lime, or ashes. Avoid adding acidifying composts like pine bark, pine needles, conifer sawdust, or grow mulch from commercial sources in the Northwest.

WATER

Water quality is a very important aspect of gardening and farming. As we know, being in a drought during the past 5 years in California, availability is a main concern when growing agricultural crops. We must have an available water source for our gardens and farms by way of agricultural ditches, wells, public water supplies, and more, but as a small farmer or gardener the quality of our water source is most important.

Many of the same principles apply to water as they do to soils. We want to find out our pH level: Is it acidic or alkaline? Is it a hard water or soft water? We've heard the terms, but what do they mean? Hard waters are usually alkaline waters with high pH that contain high amounts of calcium and magnesium and many other nutrients and minerals. It's the water you see that leaves the white residue in your shower or on your car. These waters are common in the Great Basin of Nevada and Central Valley and most wells throughout California. In general, they are good for agricultural operations. Soft waters tend to be pure mountain spring water; they are low in pH, with minimal amounts of calcium and magnesium and few nutrients or minerals. With these waters, many times we have to supplement minerals and nutrients into our soils to offset the deficiencies. Water tests are a little more expensive, costing up to $100; they are sent to specialized labs. Check your UC Extension Office for labs in your area. In the foothill areas especially (in the "red dirt" areas), high amounts of iron is a problem in your water, as well as high amounts of boron and arsenic. County Environmental Health Departments can do the arsenic and coliform tests for you. Your county agricultural advisor can help you analyze both soil and water tests and give recommendations for the crops that you are growing.

NUTRIENTS

In this section I focus on the different ways to add nutrients to your plants, emphasizing organic principles. I discuss what the nutrient elements are and how they are provided through composts, manures, cover cropping, crop rotation, and, finally, by adding manufactured fertilizers.

In all there are thirteen essential nutrients required for plant life. The macronutrients plants need to survive are nitrogen (for growth), phosphorous (for flower and fruit development), and potassium (for vigor, hardiness, and disease–pest resistance; essentially, developing the plant's immune system). In nature, nitrogen is provided by mychorrizal or microbial activity, and decomposition of forest litter to create composts and humus, available to plants for growth; phosphorus is provided by the weathering of rock minerals that dissolve into solution after the freeze-thaw and wind-rain erosion that then makes them available to plants in the warming soils of the spring and summer; and potassium occurs naturally from all of these processes and is overly abundant in California soils because of our temperate climate, and it's not leached from the soil. The secondary nutrients are calcium (for nutrient absorption); magnesium (for photosynthesis and nutrient absorption) and sulfur (for vigor and strength); and last the minor or micronutrients are iron, molybdenum, copper, zinc, boron, chlorine, and manganese.

> Air circulation, moisture, and temperature are the key to making a very efficient compost pile.

Plants in general use different amounts of nutrients. Some are heavy users and others are more efficient and use less. Vegetables are users of large amounts of nutrients (mostly nitrogen). The heaviest users are corn, asparagus, broccoli, Brussels sprouts, peppers, spinach, cabbage, cauliflower, celery, quinoa, tomatoes, melons, lettuces, onions, cucumber, squash, kale, chards, okra, and eggplant. Lower users are beets, garlic, turnips, radishes, kohlrabi, horseradish, potatoes, parsnips, and carrots. It is best to rotate your crops every other year with high, then low, nutrient user groups so not to deplete your soils. (See Appendix B for my companion plants table.)

Organic Composts

Organic practices and ingredients make the best compost. So what does *organic* mean? It's pretty simple in layman's terms: products that occur naturally in nature. If it's a manufactured or modified product it can't be used, plain and simple. Examples of organic compounds are manures, composts, cover crops, mined minerals (for example, gypsum, limestone, or sulfur), insecticidal soaps, biological control of pests (traps, repellents made of plants, animals, fungus, bacteria), fertilizers made with composted teas, and dried-pelleted fertilizers of animal manures and plant byproducts (seaweed, kelp, algae). Nonorganic products are synthetically manufactured products like Miracle-Gro, Roundup, Weed & Feed, and other brands. To be sure you are buying an organic product make sure it has the OMRI (Organic Materials Research Institute) label on the product; they are the watch dogs that make sure the product is truly organic. Many products will say "All Natural" or "Pesticide-Free," but the truly organic products you buy at a nursery must have the OMRI certification. The benefit of growing organically is that you are using materials that are commonly found in nature and that plants have been adapting to for thousands of years.

Composting is the process of using air, water, carbon, nitrogen, temperature, and the microbial activity of soil-borne bacteria and fungi to begin the decomposition process of organic matter to create humus. Humus is the by-product of humification of organic matter. For us gardeners, humus is the "life force" of the soil.

Air circulation, moisture, and temperature are the key to making a very efficient compost pile. Air is made up of 78 percent nitrogen, 21 percent oxygen, and 1 percent mixed gases. Organic matter is the base component of a compost pile. Organic matter can be vegetable scraps, manures, yard clippings, and leaves. The more variety you add to your compost pile, the better balance of nutrients you have in your compost. When air, moisture, and warm temperatures are in balance and present in the pile, bacteria and fungi then break down organic matter to help recycle nutrients. Composting in the Sierra does pose some challenges. In our cool climate, composting takes two to three times longer than it does in the Central Valley or the Bay Area.

Some people confuse peat moss with compost in organic gardening, but they are not the same. Compost is a by-product of plant decay, and peat moss is a type of organic matter. I don't recommend peat moss for vegetables; it is highly acidic, holds too much water during the spring and fall months, and is nonsustainable because of its harmful harvest of the native bog habitats in the Northwest. Some are turning to coconut fiber, which is great for nursery production of container plants but too expensive in my book for vegetable gardening.

Did you know that one handful of organic soil has more than 2 billion microbes working 24/7 to break down (decompose) organic compounds, many of which haven't been identified, and providing nutrients for all plants?

Some good activators for your compost pile are coffee grounds, citrus peels, and tea bags to acidify the pile, and eggshells to add a calcium/magnesium element. Carbon elements are the dry products that complete the composting cycle such as small barks, cardboard, paper, or straw. All ingredients are best if shredded to one-inch pieces or less, as it helps to shorten the time needed to create the finished compost. Don't be confused with building a worm box; that is totally different from a compost pile and has specific requirements. (A worm box is mostly used for small amounts of household waste and then can be incorporated into a compost pile along with the worms.) In general, any type of organic matter can be added to a compost pile. Because raccoons, skunks, and bears can be a problem, avoid using household scraps that contain fruits and meats. If you have to add them, puree them into a solution and pour them into the soil below the compost pile then cover with new vegetative material. It's a good idea to sprinkle some pelleted turkey or poultry fertilizer in with the pile to give some immediate nitrogen. If you do start earthworm boxes in addition to your compost piles, add the earthworm compost to your garden compost pile—they complement each other very well.

Composting structures are easy to make. The key to good composting is to keep it moist and to continue to turn the pile. A two- or three-box system works very well; construct them without flooring to have direct contact with the ground, and this will attract microorganisms from the soil and earthworms.

The first box is for all the new organic matter: layer your compost pile when you start; each layer should not be thicker than 6–8 inches:

1. Put down one layer organic matter: vegetable scraps, coffee grinds, leaves, clippings, and so on.
2. Add one layer of carbon, dry materials: shredded paper, cardboard, straw, and so on.
3. Add one layer older compost, manures, soil, or a dry fertilizer. Always top off your compost pile with some soil or compost to help trap the heat and moisture.
4. Then water it to keep moist but not wet. Allow three weeks to sit then begin to turn it on a weekly or biweekly basis.
5. You will see the pile diminish in size rapidly.

For the second box:

>Transfer the content of the first box to the second box, and continue to turn it over biweekly. Over time it will become "compost-like"; don't continue to add fresh organic matter to this pile.

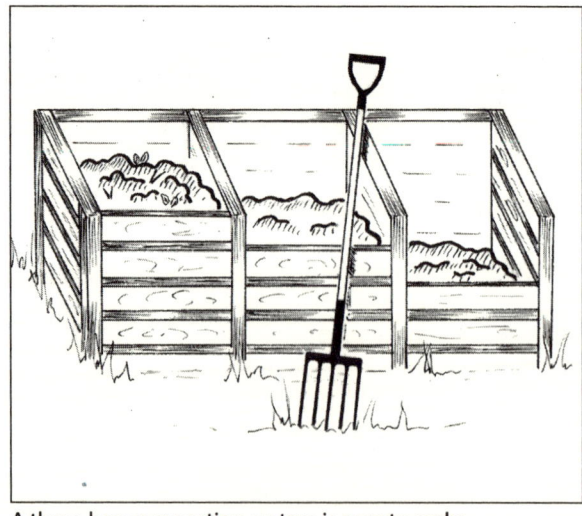

A three-box composting system is easy to make.

For the third box:
> Transfer your finished compost to the third box, and this is the one you will use for your gardens. It's a good idea to start a second or even a third succession of compost piles and let each pile compost itself, top-dressing the newer piles with the compost from the older piles. This gives you a continuing supply of compost throughout the season. If your compost pile is working right, there should not be a strong odor.

Nitrogen is very important in the process of composting, so it is essential that you add aged manures, compost, or pelleted fertilizers to your piles and turns them often. The centers of your piles must reach 150–180 degrees to kill weed seeds and diseases, but mainly to keep the microorganisms active. A compost probe thermometer makes life really easy in determining when to turn your piles. Be careful not to let the pile get too hot—if you add too much dry material it can catch on fire. That's why it's essential to not make piles more than 3–4 feet tall. For large piles, make sure you can turn them by hand or with a tractor loader frequently.

Because of the cooler climate and depending on your exposure and elevation, composting may take a good 3–4 months or even a year in some colder climates to complete its process. If you need a lot of compost for your fields or gardens, those of you in the Tahoe/Truckee/Reno area can contact Full Circle Compost in Minden/Gardenerville; they have fabulous stuff. Or check with your local nurseries and see what compost is available.

Manures

Manures are the most available and economical sources of nitrogen, minerals, and trace elements in the Sierra. In almost every rural community within the Sierra you can find some sort of livestock, for example, horses, llamas, goats, sheep, pigs, rabbits, cows, or chickens. In most instances people want to get rid of their stockpiles of manure. In organic farming fresh, raw manures cannot be used in contact with vegetables. Manures must be composted for at least 180 days before being used, or can be applied in the fall and incorporated into the soil for the next year's crop. I recommend that you stockpile all your manures throughout the summer, add them to your raised beds or fields in October, then turn them under for the winter, and incorporate 1- to 2-year-old stockpiles into your fields and beds again in spring. Composted manures (and compost teas) also work well for top-dressing your fall and perennial crops (asparagus, onions, rhubarb, horseradish, berry and fruit trees) with 2 to 3 inches around the base in late fall before winter.

Composted manures are relatively low in nitrogen and do not cause a concern for plant burn. Table 3.1 presents the general make up of most manures.

It's best to top-dress and incorporate about one pound of composted manures for every 3 to 5 feet of row, or in larger areas about 200 pounds per 1,000 square feet, or 4 tons per acre. The same rate works for compost as well. You want to keep the aged manures in the top 6–12 inches of your planter beds.

For bulb and perennial crops such as flower bulbs, potatoes, onions, garlic, asparagus, horseradish, and Jerusalem artichokes, not only are manures valuable but the addition of phosphorous inputs of bone meal and ash (potash-potassium) are beneficial to the root establishment and vigor of these crops. Remember that nitrogen can be leached through the soil over time, but phosphorous and potassium are immobile in the soil once they are placed in the soil. It is important that they be incorporated into the soil so they are available to the plant's root zone.

Table 3.1 General Make Up of Manure			
Animal	Nitrogen	Phosphorus	Potassium
Horse/cow	2%	2%	3%
Sheep/goat	1%	2%	2%
Rabbit	2.5%	1.5%	1%
Poultry	5%	2%	2%

So when is the best time to add manures, fertilizers, and composts to crops? Here are some guidelines:

- *Annuals/Vegetables*: Apply manures/compost at appropriate rates after fall harvests, then again before planting. Side-dress (add amounts of fertilizers on the sides of the plant) as needed when flowering for warm-season crops.
- *Bulbs, Tubers, and Perennial Vegetables*: Apply manures/compost in early spring, and bone meal and potash in fall.
- *Shrubs, Vines, and Trees*: Add manures/composts in early spring and early fall with bone meal and potash in fall to add organic mulches (bark, sawdust, straw, chips, and so on) as well. Keep them about a foot away from the base of the tree and concentrate out to the drip line of the tree.

- *Greenhouse Transplants and Containers*: Compost teas and commercial emulsions of seaweed and kelp are essential for growing vegetable starts and container growing in a greenhouse in an organic fertility program. A weekly and biweekly program using injectors or hose-drip tubing feeders is essential.

Fertilizers

For the purpose of this book I consider fertilizers materials that you buy from a store, organic or conventional. Most fertilizers are a make-up of natural, manufactured, or synthetic mineral products that are formulated and pelleted for ease of use through mechanical spreaders. You can buy different formulations derived for specific types of plants. Fertilizers are usually made to supply nutrients to types of plants like grasses (turf and lawn), vegetables, perennials and flowers, and shrubs and trees. Each formulation is designed specifically to meet the nutrient needs of each plant group. Like I said, the basic needs of each group are N-nitrogen; P-Phosphorous; K-Potassium (NPK), and some secondary and micronutrients like calcium, magnesium, iron, and sulfur. You can identify the nutrients on the front of a fertilizer bag. There will be a number in front of the NPK. For instance, a label will look like this: 31-4-10 (31 percent nitrogen; 4 percent phosphorous: 10 percent potassium), and it might include other percentages of minerals like calcium, magnesium, iron, or sulfur.

Next, fertilizers are classified as fast release and slow release, which is the time it takes to get the formulation of nutrients to the desired plant when water is added by dissolving it and making it available to the roots of the plants for uptake. Fast-release fertilizers dissolve quickly with water and can be poured around the plant roots like Miracle Gro products or soaked into the root ball. They are usually high in nitrogen (15–40 percent) and don't have much phosphorous or potassium. They are designed for quick, rapid growth and only provide nutrients for about 30 days. The plants react quickly to them, but because the plants can only take up so many nutrients in a small period of time the unused nutrients are then leached through the soil and end up in our waterways. These fast-release fertilizers are mostly restricted around the Lake Tahoe area and should not be used in any of our mountain areas because of their leaching properties that add nitrates to our lakes, streams, and groundwater. Most of the fast-release fertilizers like Miracle-Gro products were developed for the home lawns and commercial turf and golf course industry and are not beneficial for vegetable, shrubs, and trees because of their high nitrogen formulations and burning properties. The high salt content of these formulations will burn herbaceous plants if not watered properly. If you deprive a plant of moisture, the fertilizer salts precipitate out through the leaves during the transpiration process and the salt deposits on the leaves cause leaf burn and stress to the plant.

The second type and most beneficial fertilizer for most plants are slow-release fertilizers. The formulations release nutrients to plants at a slower level making them readily available to a plant for a longer period of time. These formulations can be organic or conventional (manufactured/synthetic) and when applied will safely provide nutrients to the plant for 3–4 months. The formulations you will find will contain lower amounts of nitrogen and higher amounts of phosphorous and potassium, more of a balanced fertilizer, which is what's needed for perennials, vegetables, shrubs, and trees. Formulations on the bag will look something like this: 8-6-4 or a 10-10-10 for their NPK percentages.

> Because our growing season is so short, it's best to grow cover crops beginning in late August–October, then turn them under for winter, so that you can start your gardens in the spring.

When selecting fertilizers for vegetables, you'll want to start with pre-plant fertilizers, which are low-nitrogen, high-phosphorous formulations, a 6-20-20, or a 0-4-4, rock phosphate, or organic bone meal, and incorporate it into your prepared seedbed. Bone meal can be found in feed stores; apply no more than 5 pounds per 100 square feet of beds. Beware that bone meal does attract dogs, so keep your source in a sealed container out of reach of your dogs. You want phosphorous to build the seedling strength of its roots whether you're direct seeding or planting transplants. Once the seedlings have 2–3 sets of leaves or your transplants have adjusted to be transplanted, usually about 2–3 weeks after planting, you can supply nitrogen formulation (as specified earlier) and incorporate it into the soil adjacent to the plants; make sure you thoroughly water them and keep the soil moist. Some good sources of organic nitrogen fertilizers are blood meal, cottonseed meal, fish meal, and hoof and horn meal. All of these are slow-releasing and will feed for up to 3 months. Apply no more than 3–5 pounds per 100 square feet of bed. Blood meal acts as a repellent against rabbits, gophers, voles, and deer if spread around the perimeter of your beds, but it also can attract bears if you're in Tahoe.

Consistent watering is the key to any fertilizer program. For heavy flowering vegetables or flowers and perennials you can add a bloom food that is high in phosphorous and potassium, and low in nitrogen during the bud stage of the plants. This will help in the flower and seed production of the plant.

Certain plants need supplemental nutrients for their special requirements; blueberries, cranberries, rhododendrons, azaleas, and conifers need acid, low-pH fertilizers, pine bark chips, and pine needles to help acidify the soil.

Cover Crops (Green Manure) and Crop Rotation

Cover crops or "green manure" are another great source of providing nutrients and organic matter to your soil. They are usually legumes (peas, vetch, clover, and alfalfa) or grass sod crops (red hard wheat, rye, bromes, oats, and wheatgrass) that are allowed to grow and build biomass (green plant weight) that are then turned into the soil by hand spading or mechanically with a disc or rototiller, where they rot, adding nitrogen, phosphorous, and potassium into the soil along with organic matter. In the Sierra, because our growing season is so short, it's best to grow cover crops beginning in late August–October, then turn them under for winter, so that you can start your gardens in the spring.

For raised beds and field operations, cover cropping or producing green manure is the most efficient way to add nitrogen and organic matter to an organic program. Cover cropping is composting in place. The idea is to build plant biomass with nitrogen fixing plants, then turn them under and let them rot over a period of time, establishing your nutrient program for the year.

In most areas in California and the foothills, up to about 2,500 feet in the Sierra, annual grassland species dominate the landscape that germinate from the rains of late fall, thus beginning the cover crop concept. Annual grains and legumes are seeded in late fall and winter with crops like oats, rye, vetches, peas, and buckwheats, and turned under in early spring to rot and be ready for planting summer vegetables and to provide nitrogen for fruit tree crops. Above 3,000 feet in the Sierra we have a different season, mostly of perennial grasslands that go dormant in late fall and rest during the winter to regrow in late spring after snowmelt, thus not getting established until summer. For those of you in the Sierra who want to cover crop, you must get your biomass and nitrogen-fixing cover crops (annuals and perennials—that is, clovers, vetches, winter wheat, rye, oats) to grow from the beginning of August through October, then mulch them and turn them under for winter to become compost for your next season. In areas above 3,000 feet it takes one season for this program versus 3–4 months in the foothills or the Central Valley. Our seasons are just too short. If you try to grow a cover crop in April or May after the snow melts and turn it under, the soils and air temperatures are too cool for

Lou Romano knows that phosphorous makes happy potatoes.

it to decay in time for the nitrogen to be available for your June crops. It will be detrimental to your crops and deplete the existing nitrogen in the soil (using it to decompose the green manure) and starve your crops of nitrogen.

Cover crops are also very valuable in protecting bare soils from soil erosion. When fields left fallow are seeded in the fall, and the cover crop is grown out into late fall, the root mass of the cover crop holds down and protects the soil from wind erosion and soil erosion from heavy winter and spring rains, along with spring snowmelt runoff. Cover crops are usually broadcast seeded over your raised beds and raked in, or broadcasted over fields and dragged in with screen or toothed harrow, or can be drill-seeded with a tractor.

Associated with cover crops is crop rotation, whereby no two crops are planted in the same plot or field longer than 3 years in a row. The objective is to take one field or plot out of production for a season to add a cover crop and allow that plot to rest and restore its nutrient base. This rest period allows the soils micro-organisms to continue to break down organic matter and turn it into nutrients available to next season's crop, along with promoting soil structure and the friability of the soil. Potatoes and broccoli are a good example of crops that shouldn't be planted in the same spot more than 3 years in a row without a rotation. Potatoes over time will develop scab (blight problem) if planted more than 3 years in a row in the same plot. The same goes for brassicas like broccoli and cabbage, which can develop clubroot, a fungal disease that attacks the crown of the stem and root causing die-offs.

Mulches

When growing gardens in the Sierra, you the gardener, master of your domain, must strongly consider using mulches in your gardens to help preserve and enhance your nutrient program. Mulches are very plentiful all around you. They come in the form of pine needles, wood chips, forest leaf molds (decaying leaves, twigs, humus on the forest floor), and vegetative mulches of straw, hay, and weathered manures. Mulches can be used for weed control, winter insulation of roots, holding moisture around plants, and dust control. They also decompose throughout the year, adding an in-place source of compost for your plants.

For winter insulation and weed control, there are a number of mulches you can use for your raised beds and walkways to reduce weeds. Straw is the best and most common. Hay is not recommended because it brings in weeds; avoid pine needles, bark chips, and those prepackaged grow mulches because they are too acidic for vegetables. Conifer mulches (pine needles, barks, chips) are great as landscape mulches around your trees and shrubs, but keep them out of your vegetables because they are too acidic. Newspapers work great, just lay out newspaper and throw straw over them and wet them down. The wet newspaper sticks to the ground and blocks out light to prevent weeds from germinating, and the straw holds moisture and warmth. It's great because it usually biodegrades in one season (if not just throw it into your compost pile), and you just lay down new paper for the next crop. Plastic mulches are popular in the strawberry industry, but I don't like to use plastic mulches because they aren't part of a sustainable vegetable program, in my opinion.

CHAPTER 4
ESTABLISHING YOUR GARDEN

To establish a garden or small farm, it takes a combination of selecting the right site and the right crops and species for your site, along with learning proper planting techniques, understanding the water needs of your crops, and knowing what it takes to sustain your gardens.

SITE SELECTION: LOCATING YOUR GARDEN

It is very important that you spend time evaluating your property to find the best possible spot to locate your gardens. You want to find flat, open land that has a nice sandy-clay-loam soil, in full sun and with as much sun exposure (8–10 hours) as possible. Try to keep away from the shading of trees and taller shrubs as much as possible. Everything else comes second. That's not always easy to find in the Sierra. When starting out, you can take soil and water samples from your property to see where the best piece of ground is and what type of water you are dealing with. You want to select a south- or southeast-facing site because that will give you the best morning sun and even sunshine throughout the day. Western sites create hot afternoon heat that can bake a garden. Try to avoid east- and north-facing sites for vegetables because they are the shady spots of the yard and don't receive much sun; they are good for spinach and lettuces for the summer, though. North-facing sites are ideal for situating an orchard in the Sierra because they create a late blooming season for your fruit trees so that they will avoid the early summer June frosts.

Planning your planter beds and laying out your gardens is very important because of the intensive nature of growing crops in a confined area. Try to centralize your irrigation water within the vicinity of your raised beds or gardens, and lay out your pathways for easy access to and from the gardens. Make them wide enough to accommodate wheelbarrows and your small equipment—for example, garden tractors and rototillers.

I think it is best to set up your vegetable beds separately for warm-season vegetables and cool-season vegetables. For warm-season vegetables like squash, corn, and tomatoes, construct raised beds at least 4 × 8 feet and 12–16 inches high with a good south or southeast exposure, and for your cool-season vegetables like lettuce, spinach, radish, kales, and broccoli, I suggest growing them directly into ground that has been amended at least 10–14 inches, also in at least 4 × 8 foot beds. Try to situate them with morning sun and afternoon shade, or with north- or west-facing exposure to keep them cooler. Remember that raising your beds for warm-season crops allows full sun exposure that will heat up those beds faster than if planted at ground level. Also, lining your

JULY & WINTER

raised beds with rock or painting your boards black will also absorb the sun's temperature and hold it longer during the evening hours.

🌰 For those of you fighting the elements in Tahoe, raised beds are a must; you will have to incorporate a mini-greenhouse over them to protect your crops from the sometimes monthly frosts. Here are some basic steps to take when building and establishing your planter beds and gardens:

1. Draw out your garden and beds to scale in feet, and locate all of the beds, pathways, water hydrants, faucets, and so on within the garden. Lay out your gardens and plan for succession plantings of the same short-cool-season crops every three weeks or so.
2. Identify what you would like to grow in each bed and whether you plan on seeding seeds, or transplanting plants.
3. Select companion plants for each bed that have the same time of maturity and water requirements. (See Appendix B.)
4. Plan for tall plants like corn or trellis beans at the north end of your beds so they don't shade your garden.
5. Construct the frame or border of your beds. They should not be wider than 5 feet across so that you can reach the middle from either side. In the Sierra you can use any type of material because things don't decompose very fast in our cool climate. Forest logs work great, any type of 4 × 6 or 6 × 8 lumber works well. Rocks are the best for the border, because they absorb heat and last forever. Try not to use railroad ties or pretreated lumber because of toxicity to the soil, and they are not certified for organic use.
6. If you are in an area of rodents (gophers, voles, moles, chipmunks, squirrels, rabbits, and so on), you will probably have to trench down to line the edges of your beds with quarter inch hardware cloth, 10–12 inches down and elbowed 6 inches outward from the beds in an L-shape to protect against burrowing rodents.
7. Consider the length of your growing season. If you plan on growing warm-season crops above 5,000 feet 🌰 or extending your season by covering your beds 🌰, it's best to add half-inch conduit hoops to your beds during construction. (See Chapter 7.)
8. If planting on a slope, terrace your hillside horizontally and plant parallel to the slope.
9. Add your amendments, and turn over the beds and get them ready for planting.
10. Once your beds are created, add your main lines of irrigation; at least one water valve for each bed.

Photos courtesy of SunMie Won.

Truckee food grower SunMie Won uses homemade cold frames to grow food year-round.

For more information, please refer to *Raised-Bed Vegetable Gardening Made Simple* by Raymond Nones or *Building Projects for Backyard Farmers and Home Gardeners* by Chris Gleason.

A raised bed at Sierra Valley Farms. The half-inch conduit hoops help extend the season.

Take advice from a good friend of mine, SunMie Won, who moved to Truckee in 2000 and was struggling to grow a garden there until she built her first cold frame in 2009, a mini-greenhouse built over her raised beds, to block out the wind and give frost protection. The beds were built with many different types of wood scraps and stones. In the spring, the soil in these mini-greenhouses warmed up the raised beds, allowing her to seed her beds in May instead of late June.

SunMie now has about ten different 4 × 8 foot raised beds with hoop frames over them scattered around her yard facing in different directions, each growing different vegetables . . . living proof that you can grow a great garden in the Truckee/Tahoe area. SunMie says that she and her kids spend 4–6 hours a week planting, weeding, and harvesting from their gardens. In addition she has added chickens to help with the household food scraps, and adds their manure to her compost program, which later is incorporated into her raised beds. Not to mention having a weekly supply of fresh eggs for her family.

The advantage of using raised beds is that you can raise more vegetable in a confined space than you could in an open field. Planting your rows close together utilizes precious space and reduces weeding. Once your beds are laid out and constructed it's time to build your soil. If you're starting a vegetable bed it's best to utilize 30 percent native material, 40 percent compost and manure blends, and 30 percent loam topsoil if available. Contact your local nursery, or if you have neighbors that have had animal corrals for years or are located in agricultural areas get a pickup load or two from them. Dig and turn over your beds at least 10–14 inches deep and remove all large rocks over 1 inch in diameter. Level and tamp. If it's an established bed, it's best to add your cover crop seed in late fall, or add your fresh manures and compost materials (leaves, clippings, compost, and manures) and let them break down over winter and incorporate them into your beds in the spring. You want to try and remove all rocks larger than 2 inches in diameter in your raised bed or field, down to about 12 inches deep in the soil. For field operations you can use a spring or toothed harrow to bring rocks to the surface and hand pick them, or a rock picker attachment that will comb your soil of all rocks 2 inches in diameter. You are now ready to plant your beds and plots.

WHAT DO I WANT TO GROW?

It's now time to decide what to grow. Vegetables, berries, fruit tree crops, or all of the above? Where will they do best on your property? Vegetables prefer a neutral-to-alkaline loam soil, relatively free of rocks, in full sun, and a flat or slightly sloped surface. Select varieties that will mature in a short season from 50 to 80 days.

Cool-season crops like brassicas and lettuce do well at altitude.

🌰 If you're above 4,000 feet you must consider growing cool-season crops (lettuces, herbs, spinach, kales, peas, brassicas, coles, Asian vegetables, mustards, root crops), or perennial crops like horseradish, asparagus, Jerusalem artichokes, and rhubarb outdoors. Consider cold frames and hoop houses for your warm-season crops (squashes, cucurbits, tomatoes, eggplant, peppers, basil, beans, and vegetable starts) to protect them from possible monthly or early autumn frosts. (See Chapters 5 and 6 for details on what grows best in our region.)

🌰 For berries and fruit trees, select north- or west-facing slopes, elevated off the meadow and valley floors, away from prevailing winds, to prevent

June frosts that can kill the buds and flowers of your tree or berry orchards. You want your fruiting berries and fruit trees to bloom as late as possible to prevent frost damage. Select rocky, gravelly, well-drained sites with good sun exposure and even canyons where specific microclimates defeat the odds.

Microclimates can work for those at altitude, too. For example, here in Plumas County, which is a very cold area at 5,800 feet, a family on the way to Lake Davis has a 3-acre orchard in a small canyon that can grow incredible apples, pears, three varieties of cherries, apricots, raspberries, squash, and tomatoes because they are protected from the harsh weather elements. Their north and west exposures keep them protected in this small canyon.

> One of the hardest things to decide is whether to direct-seed your vegetables or transplant them.

My farm (Sierra Valley Farms, at 4,980-foot elevation) is about 7 miles away in the low-lying Sierra Valley, and I can't even grow apples or warm-season vegetables because of our short season and cold temperatures settling in the valley floor. I have Grandma's old Harrelson apple tree and I get apples about once every eight years, because the frost and cold June winds kill the buds and flowers. Every year, the tree looks beautiful, but most times, there is no fruit.

Again, select berries and tree varieties that are adapted to USDA Hardiness Zones 5–7, depending on your elevation, and look for these microclimates within your property to place your specific crops. Many times semi-dwarf or dwarf trees, because they are shorter and have shorter robust branches, are better for higher altitude areas and areas of heavy snow because they are less susceptible to strong winds and heavy snows that can break branches of the taller varieties.

Should I Direct-Seed or Transplant?

One of the hardest things to decide is whether to direct-seed your vegetables or transplant them. In the Sierra, for cool-season crops, give them a chance to adapt to the Sierra's wild ride. When the soil temperature reaches 50 degrees (April 30–May 15) I like to direct-seed most of my cool-season crops like lettuces, spinach, kales, arugula, mizunas, carrots, onions, radishes, turnips, into my field; they will germinate no matter what Mother Sierra throws at them (rain, snow, sleet); they will adjust and become hardy against the elements. Plants adapt to their surroundings in nature, and cold-season vegetables are no different. Remember, once you cover your vegetables or grow vegetable starts in a greenhouse or hoop house they become adapted to the "warm-and-fuzzy feeling." Their cell walls become thinner and less tolerant of cold, so if you forget to cover one night . . . *bam!* You have lost your crop. 🌰 In general, the Sierra are too harsh to put out vegetable starts like lettuces, kales, spinach, and other leafy greens. The cold, dry climate, the winds, and the 40–50-degree daily temperature swings are too much for transplants to handle. They go into shock, and most times don't recover or take even longer to get established than if you direct seeded them in the first place. Remember you are a member of the mile-high club—your gardens and farms are 1 mile closer to the sun than those of Sacramento farmers, and the intensity of an 80–90-degree day is many times more intense than in the Valley. The crops I like to transplant are the longer-season cool crops like broccoli, Brussels sprouts, broccoli raab, cauliflower, kohlrabi, celery, herbs, and cabbage. Warm-season crops like tomatoes, peppers,and eggplant are fine transplanted, but squashes, melons, pumpkins, corn, and cucumbers all adapt better from seed.

Planting

When planting, either by seed or with transplants, the first step is preparing the seedbed. The preparation of the seedbed should be done right before planting. It's best to water the bed lightly a few days before to moisten the soil because you never want to seed or transplant vegetable starts into a dry bed; dry soils repel water, a natural phenomenon called *hydrophobic* (soils that have no polarity, positive and negative ions) and thus don't attract water, very similar to desert soils. Water will bead up on air-dry soils and act as a water repellent. Damp soils like loams are attracted to water molecules and will hydrate rapidly, allowing the absorption of water with mineral soil particles. Once the soil is moist you can then turn your beds and rake over the whole bed to remove large dirt clots or any rocks. Rake your beds level; your seedbed should have a fine layer of soil on the top 1 inch of your beds for seeding or transplanting. You are now ready to plant.

There are a number of ways to plant your raised beds and gardens, but I try to keep planted seeds together and transplanted plants together because seeds and transplants have different water requirements. Decide which beds or parts of your garden plots you plan on seeding and which ones you will be transplanting plants into and separate them according to cold-season vegetable plots and warm-season vegetable plots.

There are two theories of seeding your plots. One is to broadcast seeds across the sectioned areas of your beds and rake them in to make concentrated beds to reduce weed competition, and irrigate with microsprinklers. This is done a lot with lettuces and different greens, mizuna, or spinaches, which works fine. In some instances where you have heavy clay areas, seeds should be soaked in the seedbed and not overhead watered if possible. The pounding of water droplets on a seeded bed causes compaction of many of the Sierra clay particles that act as a "cement cap" on top of the seeds, prohibiting germination.

I prefer to seed my vegetable in rows. Never plant a seed deeper than three times its width. Too many times people plant carrots or lettuces down about 1 inch deep . . . it's not going to come up. For small seeds like lettuces, carrots, kales, and arugula up to midsized chards, beets, radishes, mizuna, and spinach, you don't want to plant them more than one-quarter inch deep. You can buy all kinds of seeders, but if you just have a couple of beds, just flip over your steel rake and make a quarter-inch depression in your soil and create a row proceeding in a straight line down the bed. Space each of these depressed rows 4–6 inches apart, then seed your vegetables, spacing them according to the seed packet specifications. Then go back and lightly back-drag the soil over the seeded row and lightly tamp. There are a number of seeders on the market; each work well in these situations. It's best to refer to different seed catalogs on what seeders are recommended for the type of vegetables that you want to grow.

Irrigation

Setting up your irrigation for watering your beds is an extremely important aspect of a successful garden. It's very important to put the water where the plants can get it and not overwater the bare ground around your gardens because weeds grow where water flows. Designing an efficient irrigation system is a must for all gardeners. For vegetables, it is far better to soak the seed with drip tubing emitters or T-Tape (see page 191), using even spaced emitters that give a uniform soak to your seeds, than to use overhead sprinklers. You can then lay your T-Tape between two rows, and like a soaker hose it will water each row on either side of the drip tubing. Refer to the www.dripworks.com catalog or You Tube on how to install T-Tape kits.

Drip irrigation was developed in Israel in the 1980s to irrigate melons. T-Tape, or drip tape, is a wonderful invention that ranges from 6 to 15 millimeter plastic tubing that has "slits" in the tubing every 6–8 inches that allows a small bubble of water to trickle out. It runs on low water pressure (10 psi [pounds per square inch]), and a garden hose can pretty much water 25 rows, 100 feet long. It comes in a roll ranging from 500 feet to over 6,000 feet and can be used for 2 to 3 years.

For larger seeds and transplants you can use the T-Tape system, or you can furrow irrigate them, which is the way my Italian grandparents taught me to set up planter beds. Furrow irrigation is a great way to utilize side dressings of manures and compost teas. In this method, hoe a furrow (grooved channel) along the length of the bed, spacing your rows 12 inches apart, with the furrow 3–5 inches deep. Then plant your larger seeds like beans, corn, or transplants of tomatoes, peppers, or eggplant on the top of the row on both sides, offsetting each other 12–18 inches apart depending on the vegetable. You then use a bubbler on the end of your hose to flood the furrow three-quarters full; do not to allow it to overflow, just let it soak in. You then water as needed, usually one or two times per week checking for moisture periodically. After about 2 weeks you can pour compost teas, spread chicken or horse manure or pelleted fertilizer and soak them in without any danger of them coming into contact with the upper leaves of the plant. After watering you can also mulch your furrows with compost or straw to hold moisture or depress weeds, but be careful in areas that have meadow mice, slugs, or earwigs—you could be asking for trouble, because you might create nesting habitat for rodents.

Another method of irrigating is to use microsprinklers (small sprinklers that run off your drip systems) that work well with seeding or transplants; however, most larger warm-season plants should not be overhead watered because it can cause mildew later in the season. Overhead watering is fine for cool-season crops. Transplants can be overhead watered to help cool them and add moisture to their leaves so that they don't dry out; this should be done a couple of times a day after planting until they go through their "shock" period.

The shock period is an adjustment period that transplants go through from being planted in a nursery pot and moved to your raised beds. They must adjust to your climate conditions of wind, sun, temperature variations of cool mornings to hot afternoons, and insects and critters. Microsprinklers do help control the problem of flea beetles on radishes and arugula and discourage aphids as well. Please refer to the catalogs of DripWorks (www.dripworks.com) for selecting from the many different types of microsprinklers that apply to your situation and how to install them.

How Much Do I Water?

This is the tricky part. Now that you've installed drip systems, T-Tape, microsprinklers, or some flood irrigating of furrows, how much water should you apply and for how long and how frequently? Irrigating your plots or planter beds depends mostly on (1) the soil type; (2) the crop you are growing; (3) the time of the year; and (4) the type of irrigation system you are using. The main idea is to maintain a steady supply of water to the plants. Letting the soil get too dry and allowing the plant to stress and wilt is not good, yet overwatering soil to the point of oversaturation depletes the air supply and slows growth, which is even worse. You need to find a happy medium, and trial-and-error is the best way to find out. Every garden is different, and every plot within the garden may be different as well. You just have to stick your finger into the soil and see if it's moist, super-wet, or dry. There are so many factors to consider when growing crops in the Sierra that you just have to be in tune with your fields or gardens and be on top of what's going on currently or what the weather may be doing in just a matter of hours. Many vegetables will fool you. Tomatoes and squash plants are a good example: on a hot day they may look wilted, or stressed for water, but you check the soil and they are still moist or wet; these are the few plants that "wilt on purpose" to save water, because the wide leaves and large number of leaves would transpire large amounts of water on a hot day and die—they are conserving energy.

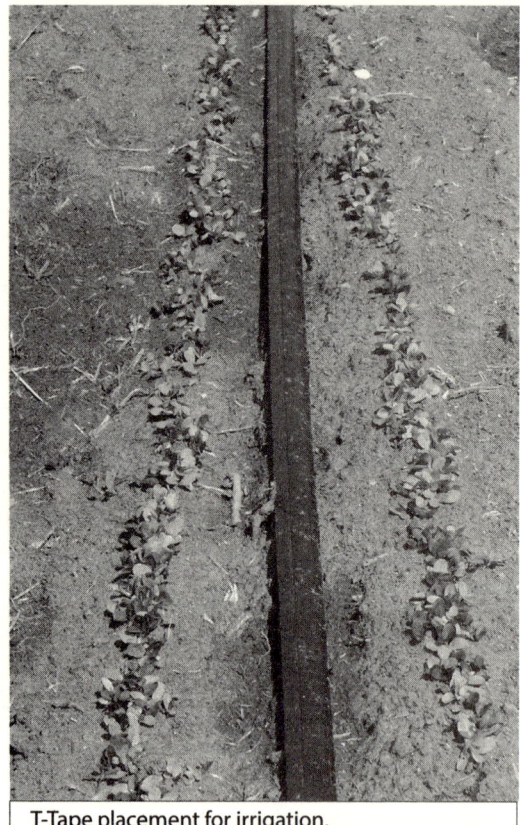
T-Tape placement for irrigation.

Always look at your soils to determine your soil structure. Do I have clay-loam, sandy-clay loam, silty-sandy loam? Do my soils hold water or not? How many days can I go without watering? The more sand you have in your soil the faster the water will percolate through the soil, while the more clay and silts you have the longer moisture will stay in the soil. You have to water and see how long it takes for your soils to dry out. That will determine your water frequency.

Consider the crop you are growing. Leafy greens are 90 percent water; so you will need to pump the water to them; don't let them dry out. I prefer drip irrigation on all my crops because it keeps the leaves clean, and I can water during the hot parts of the day without the danger of burning the leaves. I only use overhead watering for established plants of broccoli, cabbage, cauliflower, and Brussels sprouts, to cool them and help repel aphids, or in the

Close up and long view of T-Tape system at Sierra Valley Farms.

early morning for frost protection. For fruiting vegetables (tomatoes, peppers, eggplants), trellis (beans, peas), wide-leaf (squash or zucchini, cucumbers, pumpkins), or tall crops (corn, sunflowers, flowers), avoid overhead watering because they will split and discolor peppers and tomatoes, bring in powdery mildew to wide-leaf vegetables and trellis plants and knock over tall crops like corn, sunflower, and other cut-flower crops.

Overhead watering does serve a purpose in the Sierra. Don't be afraid to overhead water during the event of frosts. The ice-making effect of freezing water actually creates energy that helps all plants from freezing. Also, flood irrigating furrows or turning drips or microsprinklers among the vegetables will create a warm condition under the canopy of the crops you are trying to protect. The worst thing is to have a dry environment and freeze-dry your crops.

In general, for vegetables a rule of thumb is to add 1–2 inches of water per week to the soil surface for minimal watering; for fruit trees and berries you may have to add up to 6–10 inches of water per application to reach at least 6 feet down to the root zone of most trees and slightly less (3–4 feet) for berries.

You can use all the moisture meters you want, but the best way to assess your water needs is to be out there and feel the soil. Like I said earlier, you want the soil moist most of the time. Consistency is the best. Deep watering is preferred, instead of a shallow sprinkle. Remember, roots follow water, so if you just sprinkle with a hand hose, that's what you get: a sprinkle. A slow soak is what you want, then let it dry out to the point of having just some moisture, then do it again. Don't keep it too wet one week, and too dry the other because you didn't have time to water, and wonder why your tomatoes are cracking or you have blossom-end rot.

Be consistent and pay attention to the water demands of your crop at different times of the year. In winter and spring you are working with moisture (high humidity of 60–100 percent); because of higher humidity, soils don't dry out as fast and thus need less frequent watering, while in summer and fall you are working away from moisture (dry humidity 0–20 percent), and the soil dries out faster, so you need to add water more frequently because of the hot daily temperatures and windy days.

During the winter (in hoop houses) or spring (outdoors) you may only need to water once a week or every 10 days. As the days get longer and the soil warms up, you will have to water more frequently and for longer periods. I find that with my silty-clay loam soils, I water my cool-season crops every 2–3 days in May and June, and daily in July, August, and the first part of September, then reduce back to every 3 days and once per week until late October. I give them about a 30-minute soak on T-Tape at each application. Just stick your finger in the soil . . . that will tell you!

Tools for the Garden

I am assuming that anyone going into farming or gardening at least knows the basics of which end of the shovel to use (handle side up), and which end of a plant is right side up (green side up, roots go into the ground). With that in mind, I start with the basic hand tools that you should have in your tool shed and become efficient at using. These are the most common tools needed, along with all the screwdrivers, hammers, wrenches, compressors, and specialty power tools needed in your shop for your operation. See Table 4.1.

In general, hand tools are best purchased used from garage sales and flea markets—the older ones seem better made than the ones made today. You can usually pick them up cheap. Don't go spending $30 on a new shovel.

Table 4.1 Common Gardening Tools	
Tool	Purpose
Shovels/long handle	
Round-point shovel	Used to move or turn over soil in cultivation, to plant trees, and dig holes
Square-point shovel	Used to move loose gravel, sand, and soil
Trenching shovel	Narrow-width shovels used to clean out trenches for irrigation and power lines
Grain/snow shovel	Wide, deep-basin scoop shovels, used to remove snow or scoop grains or lightweight organic amendments like sawdust, chips, perlite, potting mixes
Shovels/short handle	
Trowel	Short-handled, hand-held planting shovels used for planting vegetable transplants, landscape annuals and perennials, and bulbs
Spade	A square, narrow-width, flat shovel–like hand tool used in deep digging, turning over raised beds, loosening of tubers and rooted plants, or to remove large rocks from raised beds
	continued

Table 4.1 Common Gardening Tools (continued)	
Tool	Purpose
Forks	
Pitch fork/manure fork	Used in turning composts, and moving hay and straw, or leafy materials
Spading fork/U-fork	Used to fracture surface, open soils, incorporate organic materials into beds, and loosen and harvest potatoes, tubers, rooted plants
Broad fork	Deep-digging steel prong fork 36-48 inches wide designed to loosen soil in raised beds, or loosen deep-rooted crops like carrots, turnips, parsnips, and potatoes
Rakes	
Fan rake	Wooden or metal fan rakes used to rake leaves, debris, and finish leveling seedbeds
Landscape rake	12-inch steel to 36-inch aluminum rakes used to level soils, sand, and gravel; used in seedbed preparation to shape raised beds for transplanting and direct seeding
Hoes/Weeding Tools: *Probably your most important tools!*	
Hoes	A wideblade hoe, 2 to 6 inches across used to remove larger weeds, pull furrows in raised beds for irrigating vegetable rows, and cultivation of soil
Planting hoe	A hand-held, short handle, V-shaped hoe used for planting vegetable starts, annual or perennial flowers, tubers, or bulbs
Stirrup/hula hoe	Have U-shapes blades like a stirrup; used for weeding small weeds along and within your rows of new crop seedlings, fast and efficient
Wheel hoe	Similar to stirrup hoes, but have a wheel attachment that make it easy to weed a lot of long rows of crops efficiently

Establishing Your Garden

Table 4.1 Common Gardening Tools (continued)	
Tool	Purpose
Cultivators	
Toothed cultivator	Long- and short-handled for hand use; for larger operations there are wheel, steel, or spring-toothed cultivators for weed removal, aerating soil, incorporating amendments into your seedbeds, or removing rocks
Pruners	
Scissor-cut	From hand-held to loppers (24-inch handle to cut larger limbs), and pole pruners (used to cut limbs up in fruit trees); provide a clean, scissor-like cut for pruning vegetables, shrubs, and fruit trees
Anvil-cut	Designed to cut larger branches; the hand anvil pruners makes it easy to cut rope for tying bunches of cut flowers
Saws	
Carpenter hand saw	Cross-cut saw used for cutting lumber; most are battery or electric skill saws
Pruning saw	Curved, rough-toothed saw used to cut tree limbs, many different handles (hand, and pole), battery powered, and hydraulic driven
Reciprocating saw	Most popular these days, the all-purpose battery saw lets you change blades to cut wood, metal, and PVC pipe for irrigation
Chain saw	All sizes from 12-inch bars for pruning to 48-inch bars for cutting down trees
Fertilizer/Seed Spreaders	
Broadcaster	"Belly-grinders," or hand-held, utilizing a spinning baffle that broadcasts out metered amounts of pelleted fertilizers or seeds, used in the seeding of cover crops or to apply pre-plant and post-harvest fertilizers
	continued

Table 4.1 Common Gardening Tools (continued)	
Tool	Purpose
Drop spreader	Small, hand-pushed or -pulled wheeled box container that drops metered amounts of fertilizers or seeds below the width of the box; used for side-dressing vegetables between rows or to seed cover crops
Seeders	
Earthway	A metered vegetable seeder that lets you adjust spacing between your seeds for individual plantings
Planet Jr.	A drop seeder that allows you to do continual seeding in a row for leafy greens production
Mini seeders	A variety of small 3-6 row seeders designed for seeding raised beds for greens production of lettuces, spinaches, mizuna mixes
Tillers	
Front tine tiller	Mostly the garden walk-behind that you can rent, good only for incorporating amendments into loose soil in your raised beds or between rows
Rear tine-rototiller	Most commercial types of tillers; they range from walk-behind 5-15 horsepower (hp) rear tine tillers to tractor-drawn sliding row tillers, and solid tillers from 4 to 8 foot long. For tractor tillers used as attachments, they take a lot of PTO (power take-off); try and select 4-6 foot tillers for tractors in the 25-50 hp range.
Pick	Used for digging into hard or rocky soils, digging narrow trenches
Ax	Used for splitting wood and cutting tree roots

How Does My Garden Grow?

I use a hula hoe religiously when the first seeds come up. Ninety percent of weeds can be taken out with a hula hoe when they are a quarter inch high.

Harvesting carrots at Sierra Valley Farms.

CHAPTER 5
VEGETABLES, FLOWERS, AND HERBS

There is probably nothing more challenging or unpredictable than trying to grow vegetables, flowers, or herbs in the Sierra. There are many vegetables, flowers, and herb species that prefer to grow during different times of the year. Some prefer to grow during the cooler times of the year in fall, winter, and early spring; those are cool-season crops. Others prefer the hotter seasons of late spring, summer, and early fall; those are warm-season crops. Most vegetables, flowers, and herbaceous herbs are **annuals**, which means they germinate (seed), grow, flower, set seed, and die all in a calendar year. Some are **biennial,** which means they grow the first year, then resprout, grow, set seed, and die in the second year, and then there are the **perennials,** which live 3 years or more. These are mostly the woody herbs, flowering perennials, bulbs, tubers, and corms (short underground stems).

SELECTING VARIETIES

When considering vegetables of either cold or warm season, it's best to select the shortest season varieties available and those suggested for cooler, northern climates. There are a number of companies to choose from when selecting seed. Because Sierra Valley Farms is certified organic, we are required to find organic vegetable seed whenever possible. If the species we want is not available organically, we must source three organic seed companies and select a variety of "untreated" seed, that is, seed that isn't treated with a fungicide or chemical to prevent pests or diseases from harming them. 🌲 In the Sierra, the number one consideration when selecting seed is days to maturity. You want to select the shortest days to maturity. Try to select all cool-season vegetable varieties that mature in fewer than 60 days and all warm-season varieties that mature in fewer than 80 days, unless you are growing in a greenhouse or hoop house.

I start by buying organic and/or heirloom vegetable seed from reputable organic seed companies like Peaceful Valley Farm Supply, Johnny's Seeds, Seed Savers Exchange, Territorial Seed Company, High Mowing Seed Company, Wild Garden Seeds, Baker Seed Company, and Siskiyou Seed Company, and local nurseries within our mountain communities like the Tahoe Tree Company (Tahoe City), the Villager Nursery (Truckee), and Perennial Nursery (Tahoe Vista), to name a few. Be careful about buying from huge conglomerate seed companies because many of their varieties may be tainted with GMO (genetically modified organisms) crops that will destroy your gene pool.

Heirloom seeds are the big craze these days. What makes an heirloom? Well, there is no certification, not even a standard. It's usually described as a variety that has been around for several generations; most are at least 50 years old. Seed Savers Exchange specializes in finding heirlooms; they are usually seeds that are sent to them from families who have been using these seeds for generations in their gardens. Heirlooms are usually *open pollinated* (which means they are pollinated by bees and insects) and develop their own traits according to their surrounding environments. They can change year to year depending on the conditions they are growing in. Sometimes you get a bumper crop, sometimes you don't.

Hybrids are varieties that have been cross-pollinated with several varieties to develop consistent traits, for example, the tomato sizes are all the same, or they fruit the same time every year. Generally, these cross-pollinations have gone on for 10-plus years. Varieties change from year to year, and your best bet is to work with seed companies that are familiar with cool climates.

It's best to order fresh seed every year, or try not to store them longer than 3 years. The viability of the seed declines after 3–4 years. Keep your seed in a room at slightly cooler than room temperature, between 50 and 60 degrees in a dry environment. Do not keep them in the refrigerator or freezer, because they will absorb moisture and could possibly mold.

SEED OR TRANSPLANTS?

How do I choose? This is a tough one. Everyone has his or her preference, but I will give you my two bits, which has worked successfully for three generations. 🌰 The criteria the Romanos use is *time to harvest before the frost gets it*. The idea is, What plant variety and method of planting is going to get your crop the quickest to harvest time (seed or transplants) before a killing frost? It's you against Mother Sierra!

Many of my fellow farmers use transplants for almost everything, from lettuce production to tomatoes. You have to remember where they farm: in the best environments in California, in the lower foothills, Sacramento and Salinas Valleys, the Bay Area, and along the Pacific Coast. These are gentle climates in the spring and early summer with mild winters, very conducive to transplants. When considering transplants that are bought from nurseries in the Sierra, the transplant is often grown in a greenhouse in Southern California, "babied" for 6–8 weeks, then weaned to the outdoors and shipped to nurseries throughout the Sierra. The first thing these plants say when they get here is *"Brr . . . where's my jacket?"* Although Sierra nurseries harden off the plants before selling them, transplants go through major shock when put into our elements. They must endure wind, cold, heat, thundershowers, and hail, and sometimes they don't make it, or they are stressed to the point that they take too long to develop. If you start from seed your plants would harvest faster than using transplants. That's why I plant 90 percent of my crops from seed, so that they can adapt to the crazy climate here in Sierra Valley upon germination.

Here is my list that I recommend you grow directly from seed:

Arugula	Lettuces	Radicchio
Beets	Mizunas	Radishes
Carrots	Mustard greens	Spinach
Collards	Onions (fall)	Sunflowers
Corn	Parsnips	Swiss chard
Kale	Peas	Turnips

Here is my list of transplants, which is pretty much everything else that's not on my seed list. All herbs, and most warm-season crops like:

Eggplant	Tomatoes	Cauliflower
Melons	Zucchini	Celery
Okra	Annual flowers	Kohlrabi
Peppers	Broccoli	Perennial plants
Pumpkins	Brussels sprouts	
Squash	Cabbage	

When using transplants, I recommend that you raise your own from seed in a cold frame, hoop house, or greenhouse so that they adapt to your climatic changes, or buy them from your local nurseries that grow their own starts.

Begin looking through your catalogs after New Year's and order your seed as early as possible. Seed your flats or six-packs in your greenhouse or hoop house 6–8 weeks before you want to transplant them into your gardens. Harden them off in a carport or unheated shade area or shed for at least a week or 10 days so they acclimate to your outdoor conditions. Once your transplants have gone through "shock" (adjusting to the planting) and have started producing new leaves, you can then cut back on the watering and let them dry out a bit. Same goes for new seedlings: when they make their second and third leaves, you can let them dry out a bit. Try to avoid buying from large chain home improvement stores because their transplants are grown in Southern California and, as I mentioned earlier, aren't well adapted to our conditions. Most times they are sold out of varieties anyhow, long before we even consider putting them in because of our late-season planting times. You have to remember big box stores are done selling vegetable starts by the end of May and early June and are then focusing on perennial flowers, bedding plants, and container shrubs and trees.

WHEN DO I PLANT?

There are many different theories on when to plant your vegetables, flowers, and herbs. I plant by the moon if the weather is cooperative, but I pretty much go by when my soil temperature reaches 50–55 degrees and when the soil has dried out enough in the spring to turn the ground over for planting. Here are some guidelines for germinating soil temperatures for vegetables:

> 50–55 degrees:
> Arugula, beets, broccoli, carrots, cabbage, cauliflower, greens, herbs, kale, kohlrabi, lettuces, onions, parsnips, peas, radish, spinach, Swiss chard, turnips

> 60 degrees and above:
> Beans, corn, cucumber, eggplant, melons, okra, peppers, pumpkins, tomatoes, and zucchini/squash

Although I like to plant by the moon, I plant many successions of the same plantings (about every 2 to 3 weeks from April through September), so I usually miss the ideal planting time for the lunar cycle. If you like to plant by the moon, here are some guidelines: It's all based on the full moon and the new moon. The 2-week period following the full moon is called the "dark of the moon," and the 2-week period following the new moon is called "the light of the moon." For the period of the dark of the moon, you want to plant flowers and vegetables that bear root crops. Remember, dark relates to crops below ground like radishes, carrots, potatoes, onions, garlic, and turnips. Thus the opposite, the light of the moon, is when you plant flowers and vegetables that bear crops above ground like flowers, tomatoes, corn, beans, and squash. Consult the annual *Farmer's Almanac* for details. See Appendix A for my Sierra Valley Farms Vegetable Planting Schedule.

In general, summer crops are a little tricky. I try to use June 15–20 as the last major frost and transplant my vegetables around that time. In most areas above 3,500 feet, if you don't plant warm-season crops like tomatoes, peppers, and squashes by June 15, they may not ripen before the frost of September. You may have to cover them early and late or grow them in a hoop house. Pay attention to your local weather forecast as you get into late August and just be prepared to cover your warm-season vegetables with row covers nightly when there is a danger of frost.

🌰 If you feel that you can grow warm-season crops outdoors, in general, direct-seed after frost danger: beans, corn, cucumbers, melons, squashes, and transplant starts of eggplant, peppers, pumpkins, squashes, tomatoes, flowers and herbs. If you have a greenhouse or hoop house and want to get an early start, seed your onions and herbs (except basil) in February or March, and you will have onion sets and herb plants to transplant in June. In your greenhouse or hoop house, seed your summer vegetable starts 4–6 weeks before planting them out in your fields or garden, and remember to harden them off in a cool outbuilding, garage, or shade cloth area a good week to 10 days before putting them out to Mother Sierra. You can also mulch them with some straw to help hold the heat and moisture when first planted; you don't want them to dry out and stress the starts. It's probably a good idea to protect them in your gardens with row covers, or other frost protection measures, if you are unsure of the nightly temperatures, and don't forget the critters . . . protect your investments from them.

HOW TO GROW VEGETABLES

Here are some specific suggestions on how to grow classifications of vegetables according to their plant parts: leaves, flowers, fruits, and roots.

Vegetable Greens

Whether you are a home gardener or a small farmer, a successful garden in the Sierra should grow successive plantings of cool-season crops (arugula, Asian greens like bok choy, beets, carrots, celery, chives, green onions, kale, leeks, lettuces, mizuna greens, parsnips, radishes, spinach, Swiss chard, or turnips). During the past 15 years here at Sierra Valley Farms (4,950 feet), we have been farming cool-season crops outdoors with no row covers from the end of April until the last harvests of carrots at the end of November. Yes, it's possible to have a 6-month growing season in the Sierra. Frost and all. Again, as a reminder, direct-seed all cool-season crops other than broccoli, Brussels sprouts, cabbage, kohlrabi, and herbs to allow them to adapt to Mother Sierra's climates and microclimates. Transplants of lettuces in the Sierra are very difficult because they are subjected to wind, sun, rain, and even hail.

Cabbage 🌰 🏺

I begin seeding them in plug trays (200 per tray, 1-inch × 1.5-inch cells) in my greenhouse at the beginning of April, and then harden them off in my hoop house for a week before transplanting them out in the field in late May, where I space individual plants 6–8 inches apart in rows. Apply a pre-plant fertilizer and side dress when they are flowering, and overhead water to keep them cool. Allow red/green and savoy cabbages to develop and become hard as a baseball, then harvest. If you leave them too long in the field they will implode, so harvest when mature and keep in cold storage 32–34 degrees.

Kale

I seed kale like spring mix, every three weeks from April until the end of August and harvest the small leaves when they are 4–7 inches in length; they are nice and tender for salads and kale chips. You can also harvest the larger leaves from individual transplanted starts when they are 10–12 inches tall. Kale can be harvested well into winter if protected, but the plants do become woody and more bitter the older they get. Store at 32–34 degrees.

Leafy Greens/Head Lettuces

There are generally two types of greens: *Leafy greens* are grown like spring mix and we harvest young individual leaves to be used in salad mixes; or *head lettuces* that are usually transplanted from starts, and are grown out as individual plants spaced 6–8 inches apart in rows; the whole head is harvested when mature. It is best to grow leaf lettuces in the Sierra versus head lettuces. (Head lettuces, like iceberg and the "ball" lettuces, will not make a dense head because our days of the season are too short and the lettuces don't develop right.) When directly seeded, the young seedlings adapt to the daily temperature fluctuations and develop thicker cell walls to cope with the wind, rain, sun exposure, and heat. In case of a frost, you may lose that one cutting of leaf lettuce, and if you cut that back, within 5 days you will have another crop. For head lettuce, once the main head is damaged your crop is ruined. In 15 years of growing leaf lettuce here I've never lost a full crop of lettuce; it may have been slowed down by frost, or lost a cutting, but never a crop. That's pretty good for one of the coldest spots in the Sierra.

I direct-seed all of my leafy greens outdoors without row covers from the end of April and succession plant every 14 to 21 days until the last week in August. I plant with an old Planet Jr. drop seeder, planting the seeds as shallow as possible (planting depth of 1/8 to 1/4 inch) to get the fastest germination possible to beat the weeds.

Leaf lettuces/spinach/arugula/mizuna/collards are harvested when young leaves are 4–7 inches tall; and cut 1 inch above the crown. They will regrow (cut and come-again), and you should get three or four cuttings in fall and early spring, and one or two cuttings from November to April in hoop houses.

Head lettuces that are grown out as individual plants are harvested as individual heads at ground level. Iceberg or romaine lettuce, bok choy, or cabbage are all done this way. All lettuces should be stored at 32–34 degrees.

Harvesting leafy greens at Sierra Valley Farms.

Swiss Chard

You can transplant this out in the field and allow plants to develop 10–12 inches tall and select individual leaves, or direct-seed them and harvest 6–8 inch leaves as a small- to midsized leaf for salads and salad mixes. You can side dress them with nitrogen 2 to 3 weeks after planting. Select individual leaves 1–12 inches tall for harvest; bunch the individual leaves in 6–8 stems, store at 32–34 degrees.

Flowering Crops

Flowering crops are the floral parts of vegetables that we eat, the flower buds; they develop usually in late fall or early spring. These crops are artichokes, asparagus, broccoli, Brussels sprouts, cauliflower, and rapini. Harvest the flower buds before they flower, when the heads are tight with tiny buds (broccoli, rapini, and cauliflower).

Asparagus

Asparagus is a long-term perennial crop. It can produce spears for up to 20 years or more. Crowns should be planted into a hill or mound in the spring 6 to 8 inches deep. Add bone meal or pre-plant fertilizer when planting. Allow the spears to emerge 12–14 inches tall and cut them off where they snap; check daily because they will grow 4–6 inches in a day and will persist for 4–8 weeks. Discontinue harvest when spears become smaller than the width of a pencil, allow them to go to fern, then cut ferns in late fall and top dress with manure. The spears are killed easily by frosts below 32 degrees, and they must be covered or grown in a hoop house at the higher elevations.

Artichokes

It is possible to grow artichokes outdoors at the lower 3,000-foot elevations if they are protected, and even though they are a perennial, you will get only two or three seasons out of them before they give out. Plan on replanting them every 3 years or so most times. They are possible in a hoop house at the higher elevations only with some tender loving care. Get plants in the spring, any of the Globe variety, and grow them for the season. Keep an eye on them because they will attract aphids and earwigs. Cut off the older leaves on the bottom as they die back. In the fall cut to the ground and mulch with 5–6 inches of pine needles or straw and cover with a 15-gallon nursery pot. Check under the pot periodically during the winter for nesting rodents. Remove covers in spring and allow to regrow. They can also be grown in containers and moved indoors for winter protection.

Harvest the artichoke when it reaches a diameter of about 4 inches. The leaves of the choke (flower) should be tight and full. Cut the stem just below the choke. Soak in water for about 10 minutes; artichokes are great hiding spots for earwigs. They will come out when submersed in water.

Broccoli/Brussels Sprouts/Caulifower
These crops can be grown together outdoors and are similar to cabbage. I seed them in 200-plug trays in my greenhouse the first of April, grow them out for 4 weeks, then harden them out in my hoop house and transplant them into the field in late May or June. I overhead-water to keep them cool, side-dress them with an organic balanced fertilizer when they flower, and begin harvest in late July, August, even into early September.

Broccoli raab/rapini: They can be grown like regular broccoli, but I like to direct-seed them outdoors and grow them out in rows like kale. Because of their tight spacing the small heads can be cut like a bunch of flowers and sold to restaurants or added to salad mixes.

Broccoli and rapini can be harvested in similar ways. Allow the main head to develop and harvest the head when it reaches about a 6-inch diameter; this will stimulate the production of side heads that are smaller and can be harvested weekly for about 4 weeks. Harvest the smaller heads in 2–4 inch diameter. Chill in cold water, shake off, and store at 32–34 degrees.

Brussels sprouts: Allow main stem to elongate and develop the sprouts. When sprouts become hard and firm, about the size of a golf ball, cut off stem with sprouts, dislodge sprouts from stem, chill in cold water, drain and store at 32–34 degrees.

Cauliflower/Romanesco: There are white, yellow, and purple cauliflower, and Romanesco varieties. All can be harvested the same: Harvest the main head when full at about 6–8 inches in diameter, soak in cold water, shake off, and store at 32–34 degrees.

Vegetable Fruit Crops
These are your summer crops, like beans, corn, cucumbers, eggplant, melons, okra, peas (spring), peppers, pumpkins, squash, tomatoes, and zucchini that produce "fruit" from their flowers. Each has its own time of maturity.

Most summer vegetables can be stored at 45–55 degrees in a cool garage or shed.

Beans /Peas
There are "bush-type" (grow like a bush) and "pole-type" (grow up poles) varieties of beans. Peas and beans are planted the same way. I prefer to direct-seed them, but they can be transplanted as well. Pole bean varieties produce more beans and for longer periods than bush beans. Heirloom Romano Pole Beans are the best (of course, with a name like Romano!). For peas and beans I

like to make furrows with a hoe about 4–6 inches deep and seed the bean on the top of the ridge, then flood the furrow high enough to soak the top of the ridge. Once the seedling has emerged, add your poles and wrap the tendrils, add side dressing, or compost teas into the furrow after two weeks. Water weekly till harvest. Bush beans do not need staking. Harvest bush and pole varieties when pods mature while green, clip or twist from the vines, and avoid breaking the vines. Pole varieties produce longer than bush beans. Store at 32–34 degrees.

Corn

Corn can be grown only in the right areas between 3,000- and 4,000-foot elevation on the west slope of the Sierra, Susanville, and selected areas in Reno and Carson City after the danger of frost in June. Direct-seed three kernels of corn about 1 inch deep on the top of the ridge of a furrow, spaced about 18 inches apart. Corn is a heavy nutrient user so add a balanced fertilizer or compost and manures when planting. Side-dress with nitrogen once the tassels begin. Check husks for corn worms when tassels begin to form; add a tablespoon of mineral oil to each tassel to kill eggs of corn worm (timing here is essential). Harvest corn when kernels have filled the husk to the tip and are fully formed. Snap husk quickly in a downward-twist motion to harvest. Store at 34–38 degrees.

Cucumbers

Can be grown on ground or on a trellis; clip stem when cucumber has reached its desired length. Flowers can be thinned and vines pruned to improve production. (See Eliot Coleman's *Winter Harvest Handbook* for details on growing cucumbers in a hoop house.) I plant them in my hoop houses and recommend anyone over 4,000 elevation do the same. I plant them as a transplants, grown from my greenhouse incorporating a balanced fertilizer during planting. Plant them on ridges of a furrow 12–18 inches apart. Train them to grow on poles, plastic fencing, or on elevated strings. Very prolific. Harvest every 2–3 days; crop will persist for about 4–5 weeks. Store dry at cooler room temperatures 45–55 degrees.

Eggplant

Grow under similar conditions as cucumbers and the same way as corn. Stake up your eggplants and harvest when they reach full color and size according to their variety. Use small clippers to cut the stems, or a quick twist to harvest fruit. Avoid breaking the main stems. Harvest weekly for 4–6 weeks. Store dry at 45–55 degrees.

Vegetables, Flowers, and Herbs

Melons/Cantaloupe/Honeydew/Pumpkins/Gourds

Melons and this group must be grown outdoors. They take too much room indoors. We grew them as kids in San Jose on the flower farm. Those of you in Susanville, Reno, and Fallon on the east side and 3,000 feet on the west side, hot areas in the Sierra, can get away with short season varieties. It's best to pull furrows, plant by seed or transplants halfway up the furrow, and flood irrigate. They grow well with drip irrigation as well. Avoid overhead watering; this will create mildew. When melons form a good size and turn color, cut back water, roll melon into the furrow, and allow leaves to shade fruit so they won't burn in summer heat.

There are quite a few melons out there, so be familiar with the melons you are going to raise. As the melons near maturity, the vines will die back and you can then stop watering. The stem will sometimes detach from some melons (cantaloupe), or will easily detach. It's best to leave them to ripen in the field as long as possible to become sweet. Harvest weekly as they ripen in August–September. Store dry at 45–55 degrees.

Squash, Zucchini

These are best grown on mounds with a furrow dug around them about 24 inches in diameter, about the size of a tire, two groupings of three planted seeds planted 1 inch deep and about a foot apart into the raised mound; flood irrigate to soak the mound. After emergence, keep weeded and add a side-dressing balanced fertilizer when flowering.

There are male and female flowers. If you want to harvest squash blossoms to eat, you must harvest the male ones. The female flowers produce the fruit. Most male blossoms are on the outer part of the plant, are large and more of them bloom first. The female ones are usually centered in the middle of the plant, are smaller, and have a swollen stem below the flower, which is the zucchini or squash beginning to develop. For the fruit, once fruiting starts, harvest them daily. A good twist will harvest the fruit, or use a kitchen knife. Best stored at room temperature dry at 45–55 degrees.

Peppers: Hot or Sweet

Peppers are a favorite of mine. Hoop house production only and plant similarly to tomatoes, with 12–18-inch spacing depending on the variety. They will need to be covered in fall for frost protection. Peppers like early application of a balanced fertilizer until flowering; then cut back on the water when fruiting. Drier, hotter conditions are better for peppers. Allow them to ripen to desired size and color. Harvest peppers every couple of days during harvest season for about 4–6 weeks. Clip pepper stem when harvesting. Stems are very easily broken if the peppers are not properly twisted off, or cut. Peppers should be kept dry at slightly cooler than room temperature, from 45 to 55 degrees. Never in a refrigerator.

Tomatoes

There are two types of tomatoes: *indeterminate* (which means they have no boundaries and need trellis/staking), and *determinate* (which means they have a determined size, usually in bush form). Tomatoes, especially heirlooms, are long-season vegetables that need 70 to 110 days to harvest and in most cases must be grown in hoop houses. I plant them on ridges of furrows with 24–36-inch spacing depending on the variety and side dress upon flowering. Select the shortest growing season varieties. A variety like Siberia can be grown outdoors and then covered in late August at the lower elevations around 3,000 feet. Above 4,500 feet, start your seeds in March in a hoop house or greenhouse and have them ready to transplant into your hoop house by the first of June. You will have to use row covers to protect them from June frosts. If you are planning on growing tomatoes in a hoop house for fall harvest, select determinate varieties that stay in small compact-bush form; they are easier to cover with protective cloth in August. For pollination, it's best to leave the hoop house or greenhouse doors open every day to allow a breeze and insects to enter the greenhouse, and have adequate floor ventilation. Consistent irrigation and slightly drier conditions when fruiting will produce a better tomato. Add a calcium-base fertilizer during flowering to prevent end-rot (calcium deficiency), or add crushed oyster shell in your pre-plant operations.

Tomatoes must be left to ripen on the vine for best quality. Clip stems when tomatoes are ripe, and place them in single layer flats whenever possible. Try to avoid stacking tomatoes whenever possible. Tomatoes are very perishable, especially Heirlooms; harvest ripe, or slightly under ripe, and store in single flats at room temperature 45–55 degrees. Never put tomatoes in the refrigerator. If your plants get frost damage and you have green tomatoes on the vine, cut the plant off at ground level and remove as much of the leaves as possible, leaving only the tomatoes, and lay them out in a garage or hoop house and allow them to ripen on their own. There is enough juice in the stems to allow the tomatoes to ripen. You can sprinkle them with water during midday to add moisture to the stems. I've picked ripe tomatoes using this process for a month, and made fried green tomatoes or canned them.

Vegetable Root Crops

Pretty much everything that grows underground (for example, beets, carrots, garlic, horseradish [perennial], Jerusalem artichokes, kohlrabi, onions, parsnips, potatoes, radishes, shallots, and turnips) can be grown outdoors in the Sierra beginning in April and lasting through October. For most root crops, I begin direct seeding in April for midsummer-to-fall harvests of these crops. Crops can be thinned for larger root development.

Harvest is done by hand with a shovel when the crown of the root has shown full development. For field operations, a small plow or mechanical harvest can be used. All crops when harvested are washed and placed in storage. Most root crops are best stored in a cool, root-cellar dry storage 40–50 degrees; or in cold storage 34–38 degrees.

Beets 🌰 🌲

Beets are a little temperamental in my book; they are a hit-and-miss crop. I find it best to pre-soak the seed overnight, let them air dry, then seed them about 2 seeds to an inch, about one-quarter- to one-half-inch deep. The key is to keep the soil moist until germination. Beets like warmer soils, so seed them outdoors beginning in late June, or they can be started in the hoop house and transplanted out in your bed at 3-inch spacing. They are about 60 days to harvest.

> The colder the ground gets in fall-to-winter the sweeter the carrot, parsnip, or turnip.

Carrots/Turnips/Parsnips 🌰 🌲

Carrots, turnips, and parsnips do well in deep loam soils in your raised beds. They do not like acid soils, so make sure your pH is 6.5–7.0 to start. It's best to grow the midsized carrots like a *chantenay* variety (75 days to maturity), but any 5–6-inch carrot does best in our somewhat shallow Sierra soil. I direct-seed them in May, June, July, and early August with my Planet Jr. seeder, and water them with T-Tape. Harvest begins in late July, and I have carrots until Christmas. If you want larger carrots thin them to about one per inch; if you want them smaller don't thin them at all. (I don't; I prefer the smaller ones.) The colder the ground gets in fall-to-winter the sweeter the carrot, parsnip, or turnip. The cold temperatures crystalize the starches, turning them into sugars, which makes them really sweet. The deer herds come into my fields and eat the tops off, and I can just follow behind them and harvest my carrots, also fertilizing my fields while they fatten up for winter.

Horseradish/Rhubarb/Jerusalem Artichokes 🌰 🌲

I put these together because these crops have similar growing conditions, even though horseradish and rhubarb are perennials. Jerusalem artichokes (sun chokes) act like perennials; like horseradish, once you put them in the ground you can't get rid of them! For planting horseradish, just take sections of the

Chantenay carrots from Sierra Valley Farms.

root and plant it 4–5 inches down in nice loamy soil, and you're good to go with weekly watering. The leaves will yellow in October; take out the center root 6 inches deep with a spade anytime from November to March and leave the rest of the roots in place and it will regrow year after year. The same goes for "sun chokes"—take the tubers and bury them 3–4 inches into the soil, and water weekly. They then multiply like crazy under the earth. When the tall stalks flower and begin to yellow your chokes are ready to dig up. If you leave 10–20 percent of the tubers in the ground every year you can create perennial beds; the same goes for horseradish. Rhubarb is a beautiful perennial. Find a nice protected spot and harvest the rhubarb stems in spring/summer when they turn reddish color at about 12–18 inches tall and cut off the leaves. They can be harvested green any time.

Onion/Garlics/Shallots

I plant two crops of onions or shallots of any kind, one in the spring with bulb sets and one in the fall with seed. I like to direct-seed onions in the field and get them established in August–early September, and let them go through the winter to grow out for a June–July harvest; the others I seed in plug trays in my greenhouse in February, to have sets available to plant out in May–June for a fall harvest . . . onions year-round! Same for garlic, fall planting (hardneck) and spring planting (softneck) of seed garlic (cloves) gives a full year supply. Once the leaf tops drop and the bulb is formed, water is discontinued, and as the stem dries, the onion is harvested and used fresh or stored in a dry, cool, aerated barn or garage. Store dry 45–55 degrees. Do not wash them.

Potatoes

Grandma said always plant them in the spring when the dandelions bloom. Quarter them with each having a couple of "eyes" (growing buds), dip the cut end in ashes, then plant 4–5 inches deep with the cut side down in raised beds or raised furrows. Feed until flowering.

If the plant flowering and the tops get killed by frost they will still bear fruit. For harvest once the tops begin to yellow and die back, cut off the watering, and harvest in late August to Labor Day. You can harvest a little at a time weekly by digging around the plant and pulling them off, or as I do just dig them all up and harvest the whole crop. It's best not to wash potatoes, because the moisture may cause rotting. Just wipe off dirt and place in dry storage 40–50 degrees.

Radishes

Radishes are my favorite! If you can't grow a radish then you probably shouldn't be gardening. They are the easiest to grow, at least in the spring. In some areas they do bolt when the night time temperatures get above 60

Onions seeded in plugs trays in the winter are ready for transplant (as sets) in the spring.

degrees, especially on the east slope of the Sierra, or the Reno/Carson City area. I grow them like carrots, direct-seeded outdoors every 2 weeks from May until Labor Day. All all varieties do well in the Sierra. Harvest when the radish is visible at the soil level in 40–45 days.

SEED SAVING

All of you farming or gardening in the Sierra above 3,000-foot elevation should be saving your own acclimated seeds of vegetables, flowers, and herbs. It's really pretty easy. Collecting your own seed is the best way to protect your seed bank from GMO (genetically modified organisms) contamination. If you buy seed, the only way to ensure that you are not buying a GMO-contaminated seed is to buy organic seeds.

Basically, it's species selection and survival of the fittest in the Sierra. You need to be observant about what's happening with your crops. First, it's very important that you mark down the variety of vegetable that you are planting. I use a logbook to mark down all the rows of vegetables that I plant. I mark down the field number, the date planted, the variety of seed and the seed company, and how many rows or plants that were planted. Throughout the season I make notes about different varieties, such as their vigor in growth, flavor, appearance, color, productivity, and longevity, or cold hardiness. Did they do over and above any other variety, or were they just okay?

The dominant vegetable plants that catch your eye or your attention are the ones you want to save. The same goes for flowers and herbs in the garden. Was there one plant that produced an incredible flower or a slightly different color? Same with herbs. Did one have a different look to it or have fantastic flavor over the others? Those are the ones to collect seed from, or take cuttings from if they are perennials.

The other issue is isolation. Once you decide to select your own seed you can't plant them side by side with their relatives or they will cross-pollinate, so you need to research who's related to whom . . . for example, you don't want to plant squashes, zucchini, gourds, and pumpkins together because they may cross-pollinate and inherit characteristics of the other vegetable plant. It's very important to plant selected seeds in different beds and different fields to prevent cross-pollination.

There are two ways to collect seed for vegetables. First, once you've isolated a particular plant that you would like to save, flag it with marking tape or place a colored stake by it, and log the variety and seed company that it came from. Next, as it goes into flower, keep an eye on it as it sets its seed head, usually about 3–5 weeks afterward. You need to research the vegetable you are trying to save to make sure you know what the seed heads look like. It's best

JULY & WINTER

Collect different sizes of mesh tools for seed screening, so that you have the right tool for the specific seed (top); place seeds in plastic bags only when you're sure they are dry, otherwise they will mold (bottom).

to become familiar with the plant families the vegetables are associated with because the seed capsules will all look the same. For instance, all vegetables in the mustard family, such as arugula, broccoli raab, mustard greens, radish, and turnips, are easy to distinguish by their pods.

The key to collecting seed is timing. You don't want to collect it too early when the seed is immature and not viable, but you don't want to collect it too late when it opens and you've lost the seed. First, you want to shut off the water to your plant when it's in flower. By stressing an annual plant like vegetables, it triggers a sense that in the plant that it's going to die and creates a lot of seed. There are two ways to successfully collect the seed. When the seed heads have developed, you'll see that the seed capsule or pods are showing some form in which the seed is enclosed (like a bean pod) and the plant will start to turn yellow. At this time you can either place a paper lunch sack over the head and twist-tie the bottom (this protects the seed and prevents birds, insects, and rodents from eating your seed), or you can cut or pull the plant out, bunch it, and hang it in a barn or garage to let dry. Once hung, it's best to place a paper shopping bag over the seed heads and tie them off, then with a marking pen note the date, field, or raised bed it was grown, variety, and seed company that it came from, and forget about it until around November.

What's great about collecting seed is that it's a good winter project when there is nothing else to do. Around November, you can then see how dried the stems are sticking out of the bag. Gently tap the stem, to a slight shaking, and you will hear the seeds fall into the shopping bag. Take them into a garage or a comfortable working shed and lay a tarp down to begin working the seeds out of the bags. Try to manually dislodge most of the seed from the stems and remove all the stems and twigs. Just leave the seed and husks in the bag. Work one species at a time so that you don't mix up the varieties. Next it is time to screen. It's best to accumulate different meshes of screens from one-quarter-inch hardware cloth down to the fine screens. Select a screen that allows at least 80–90 percent of your seed to easily slip through. Pour out the seed from the shopping bag into a small clean bucket or large bowl. Manually dislodge all the seeds from the existing pods and capsules and discard as much small twigs and chaff as possible. Place the selected screen over a bucket or bowl, and pour the seeds gently, stirring them around until most of the seeds have fallen through. Discard the chaff residue and do it several times. Your seed should be pretty clean.

The last step is blowing out the final small chaff; this can be done on the porch, or a protected area that has a slight breeze or with a small fan. You don't want to do this in a gale-wind area like Sierra Valley, or there goes your seed. Take your bucket of cleaned seed, and hold it about 1 foot over another bucket and pour the seed slowly, allowing the breeze to blow any loose chaff away. This can be done several times until you feel the seed is clean enough. You can also do this with a small fan, which is what I do in a garage or barn, because the winds in the Sierra are always gusting and you never know when one is going to surprise you.

Once this is done, your seed collection is complete. For storage I recommend you either store them in paper lunch sacks, or if you have a lot of seed store them in Mason jars or large mayonnaise jars and punch a half a dozen small holes in the lids to let them air dry. You don't want to put your fresh seed in plastic containers, or Ziploc bags, because the moisture will cause them to mold. By the time spring comes, your seed will be dry, and then you can transfer them to Ziploc bags. Always make sure to put labels on your seed identifying when collected, date, what field or bed grown, variety, and seed company bought from.

For storage keep them in a dry place, not the refrigerator or freezer. Keep on a shelf at around room temperature, between 50 and 65 degrees. Most vegetable seeds will be viable for 1 to 2 years. It is best to at least collect seed about every 3 years if you want to save your own seed. Throw out all seeds that are more than 3 years old. This method works well for herbs, flowers, and perennials.

CHAPTER 6
FRUITS, NUTS, AND BERRIES

If you are in the right place, it is possible to have some success growing fruit trees, berries, or nut trees on your property at elevations between 3,000 feet and 4,500 feet. Above 4,500 feet to 7,000 feet it is going to be chancy, at best. The thing to remember is that you will find varieties that have rootstocks that survive down to –50, and, yes, that is correct. Those varieties will survive the Sierra, but what prevents the trees from fruiting at our higher elevations is our May–June killing frosts and cold spring winds that kill the flower buds and flowers so you don't have any fruit for that season.

Just because you are at a high elevation (say over 6,000 feet) doesn't mean you can't grow fruit trees. The key is finding that spring microclimate condition. It doesn't matter what happens the rest of the year, but you want to be in an area that has a balmy spring, say, higher morning temperatures above freezing in March through May. If you are interested in fruit trees, it's best to invest in some high-low thermometers, either digital or manual, and begin to hang them in trees or place them on a fence or selected posts and record your spring temperatures between May 1 and June 20. It is said that exposed flowering buds and flowers can withstand temperatures down to 24 degrees, and anything below that will be killed. You can purchase frost alarms that will notify you when the temperatures reach freezing (32 degrees), and then you have a few options:

If you have only a few trees you can cover them with agribond fabric, blankets, or plastic sheets. Commercially, 8–10-foot-high sprinklers can be spaced between trees and watered continually when the temperatures reach 30 degrees. Water and icing (water freezing) will create warmth as long as the ice is dripping and will help against frost damage to the plant tissue down to 27–28 degrees. If it gets colder than that and begins to ice up it will cause more damage, and can build up and break branches. Ask yourself, "What is the forecast for our predicted lows?" If it's below 25, covering is your only chance.

When selecting a site for fruit trees, remember some basics: Heat rises in the morning and declines in the evening; the sun comes up in the East and sets in the West; wooded canyons are warmer than open meadows; and dense forests are warmer than open areas around water. With this in mind, look for an elevated piece of your property or protected canyon that has full sun with northern or west exposure (you want the trees to bloom as late as possible), and preferably close to a water body (stream, pond, or reservoir). All these factors will help the success of your fruit tree selection. Look at your prevailing wind direction in the spring, especially if you are in a windy canyon or open, exposed area. You may want to establish some windbreaks a year or two before you start an orchard to help protect your trees. Check with your local Natural Resource Conservation Service (NRCS) or Resource Conservation District (RCD) for cost-sharing programs that will help you put in windbreaks.

FRUIT TREES

Fruit trees like deep, well-drained soils and do well in our Sierra acidic soils. Bees and insects must pollinate all fruit trees or they will not bear fruit. There are two types of pollinated fruit trees: (1) **Self-fruitful**, which are self-pollinators and do not need a pair to pollinate, are apricots, European plums, peaches, quince, and sour cherries. (2) **Self-unfruitful**, which need to be pollinated, are apples, pears, sweet cherry, and Japanese and American plum. In most cases when considering fruit trees it is best to plant a few different varieties to ensure pollination. If you plan on starting a fruit orchard you need to plan either to start your own beehive or bring in bees for your orchard during the bloom period. Some bee guidelines for young trees (under 8 years old) are to have two hives per acre, then decrease to one per 10 years or older, and when in full bloom, try to keep at least one-half of a hive per acre. If you have neighbors who have hives within 300 feet, you can share hives. Check with your County Agriculture Commissioner (CAC) on the locations of your local beekeepers.

Once you have selected your spot for fruit trees, the next challenge is finding the right species and varieties. In most cases, apples and pears are the hardiest for the Sierra above 4,500 feet. From 3,000 to 4,000 in selected microclimates you can get away with apricots, cherries, and a peach or two. As I mentioned earlier, there is a gentleman in Portola, up by Lake Davis, who is in a protected canyon at 5,800 feet; he has three acres of beautiful sweet and sour cherries, apples, and apricots, so it can be done.

When selecting fruit trees from mail order or a nursery, they are usually 1–2 years old. Most fruit trees should be bought bare root (not in a container) in the spring from a local nursery or online. They should have good root structure (long roots, branching and full), no molds or mildews, and be straight and healthy looking with no dead twigs. There are standard trees (full-sized), semi-dwarf, and dwarf trees. In the Sierra, each has its place. Standard trees have better root systems, live longer, and bloom later to help with the late June frosts. Dwarf trees are shorter, easier to pick, bear fruit earlier, are better in high wind areas, but bloom earlier and can be affected by June frosts. The semi-dwarf are somewhere in between. The young trees can be bought as seedlings or whips (first-year, single-trunk, unbranched), or 2-year-old branched trees that will take a few years to bear fruit.

In general, this is how long it takes for newly planted fruit trees to bear their first fruit:

Apples, apricots, sour cherry, peaches 3–5 years

Pears, plums, sweet cherry, quince   4–7 years

Bearing fruit in the Sierra can be affected by factors such as climate stress, plant health and nutrition, winter and animal damage, pruning and water stress. These figures may vary depending on your site selection.

There are many nurseries online that are you can choose from, and most are from outside California. Following are some California nurseries that I like that have heirloom and cold hardy varieties. Check them out!

(TA) Trees of Antiquity (Certified Organic & Heirloom): 20 Wellsona Road, Paso Robles, California 93446; 1-805-467-9909, or www.treesofantiquity.com

(LC) LE Cooke Company, 26333 Road 140, Visalia, California 93292; www.lecooke.com.

For local Sierra Heirlooms:

(FG) Felix Gillet Institute: P.O. Box 942, North San Juan, California 95960; www.felixgillet.org

One out-of-state nursery that I've used is Indiana Berry Company **(IB), www.indianaberry.com**, 1-800-295-2226, or info@indianaberry.com.

Harvesting and Storage

When it's time to start harvesting the fruits of your labor, timing is the key. Pick no fruit before its time! Easier said than done. It's very important to start paying attention to your fruit a couple of weeks before harvest, because different varieties and trees can start ripening before others. Most fruit in the Sierra don't ripen at all at once. Pests like birds, squirrels, and even bear will strip trees at harvest time, so you need to have a plan of action and be ready. If something can go wrong it will go wrong, so be observant and prepared.

If you pick it too soon you lose flavor and color; too late and it won't last long and become mushy. Doing a taste test is the best method: pick a fruit, cut it open, and eat it. During harvest, usually twisting a fruit and pushing forward will disconnect the fruit from the limb. The key is to keep the stem intact, and de-stem later. Avoid bruising the fruit; line some picking buckets with a clean rag or towel, and use cotton gloves to pick fruit. Take care of your product. Pick when it's cool, in the mornings or evenings, and allow the fruit to chill in an open-air barn or cool shed. The ideal condition is to store between 36 and 41 degrees at 80–85 percent humidity if you have cold storage. Apples and pears store the best; peaches, plumbs, cherries, and apricots should be picked and used within a week after picking. Daily and every-other-day picking schedules should be performed throughout the harvest; it's easy for these fleshy fruits. It's best to remove all fruit from the trees or orchards and feed to livestock or dry them. If left on the ground to rot, the fruit will attract birds, raccoons, deer, and bear to your crops.

HOW TO GROW FRUITS

Apples

Apples like well-drained soils and slightly sloped (2–4 percent) eastern exposure sites. Most apples like warm fall days and cool nights, and the apples shown here are selected for Zones 3–7. You want to select late season blooming varieties. A late season and an early season variety will not pollinate each other; they have to be blooming at the same time. They grow best with a balanced fertilizer program of composts and manures in spring and fall, and in areas below 10 degrees it's best to mulch around the rootstalks with pine bark, needles, manure, or straw for winter freeze protection.

Apples are produced on spurs, which are small fruiting branches that produce the flower buds, hence the fruit. Most apples are low maintenance and need little pruning annually to enhance fruit production. Apples usually produce more fruit than they can carry, so some fruit thinning should be done when the apples begin to appear. It's just a matter of pinching off all but one or two apples per cluster, within 4 to 6 inches from each other. Try to gingerly remove the small apple without damaging the spur (small dog-leg branch that produces the clusters). It's easiest to hold the stem between your thumb and forefinger and push, or "flick" the fruit from the stem (spur).

When selecting apple varieties, start with those that have a cold hardy rootstalk like Antanovka. They are the roots that the fruiting bodies are grafted to; called the *graft union*, which is usually about 6 inches above the crown of the root (ground level) on whips and 2-year-old transplants. Table 6.1 shows some apple varieties and the region of the Sierra in which they would do best.

Table 6.1 Suggested Apple Varieties				
Variety	Region	Bloom	Fruit	Uses
Norkent	🌰	mid-Sept	red	eat
Norland	🌰	early-Aug	red	eat
Haral Red	🌰	mid-Sept	red	eat

Table 6.1 Suggested Apple Varieties (continued)			
Variety	Region	Bloom	Fruit	Uses
Haralson	🌰	mid-Sept	red	eat
Honeycrisp	🌰	mid-Sept	yellow	eat
Honey Gold	🌰	late-Oct	yellow	eat
Liberty	🌰 🌲	late-Oct	red	eat
MacIntosh	🌰 🌲	late-Sept	red	eat
Northern Spy	🌰 🌲	late-Nov	red	eat
Spartan	🌰 🌲	late-Oct	red	eat
Freedom	🌰	mid-Sept	red	eat
Zestar	🌰	mid-Sept	red	eat
Heirloom				
Wealthy	🌰 🌲	late-Sept	red	eat
Yellow Transparent	🌰	early July	yellow	eat
Northwest Greening	🌰	mid-Sept	green	dessert
Rhode Island Greening	🌰 🌲	late-Sept	green	dessert
White Pippin	🌰	late-Sept	yellow	eat
White Winter Pearmain	🌰 🌲	late-Sept	green	cider
Christmas Pearmain	🌰	late-Sept	red	eat
Yellow Pearmain	🌰 🌲	late-Oct	yellow	dessert/cider
Fit for the Czar	🌰 🌲	late-Oct	red	storage winter
Galloway Pippin	🌰	late-Sept	yellow	cider/eat
Margil	🌰 🌲	late-Oct	red	eat/storage
Reinetter Franche	🌰	late-Sept	yellow	dessert/cook
				continued

Table 6.1 Suggested Apple Varieties (continued).				
Variety	Region	Bloom	Fruit	Uses
Organic				
Alexander	🌰	mid-Sept	red	eat
Amere D'Berthecourt	🌰	mid-Sept	yellow	cider
Bramley's Seedling	🌰	mid-Sept	red	cooking
Campfield	🌰	late-Sept	green	cider
Court Penduplat	🌰 🌲	v. late-Sept	yellow	eat
Courtland	🌰 🌲	late-Sept	red	tart/eat
Duchess of Oldenburg	🌰 🌲	early-Sept	red	pies
Fireside	🌰 🌲	late-Sept	red	eat

Apricots

Apricots, believe it or not, come from cold climates where they must bloom very quickly after their chilling requirements are met. 🌰 Because apricots are early blossoming and fruiting fruit trees, this makes them tough to produce in the Sierra, even at 3,000-foot elevation. Apricots tend to bloom in early April, and fruit is ready for harvest in late July and August. Apricot buds and flowers are very susceptible to frost, even light frosts around 29–30 degrees. Apricot rootstalks are rated for Zones 4–9. Apricots can also be grafted to plum rootstalks. The Manchurian apricot is the hardiest of all apricots and the latest blooming, and would be the one to try first. Apricots, like cherries, do not like wet clay soils and prefer a sunny eastern exposure, protected from prevailing winds. Apricots grow similarly to plums and like well-drained soils; they do need summer watering because of their early season crop production. Apricots are very perishable; taste first, then harvest when ripe. Store at room temperatures for short-term use, or place in cold storage 34 degrees for longer periods. Table 6.2 shows some apricot varieties to experiment with.

Peaches

🌰 Everybody wants peaches, but they are a problem in most areas. Table 6.3 presents peach varieties and their regions. Your best bet is to have a few trees and baby them; plan on covering them during the bloom

Table 6.2 Suggested Apricot Varieties		
Variety	*Region*	*Fruit*
Scout	🌰	yellow
Moongold	🌰	yellow
Morden 604	🌰	yellow
Harlow	🌰	yellow
Heirloom		
Harcot	🌰	700 chill hours*
Puget Gold	🌰	700 chill hours
* Chill hours is the number of frost hours a fruit tree needs to be in dormancy before it leafs out in the spring to produce fruit.		

Table 6.3 Suggested Peach Varieties		
Variety	*Region*	*Fruit*
Reliance	🌰 🌰	freestone
Canadian Harmony	🌰	freestone
Contender	🌰	freestone
Empress	🌰	freestone
Glohaven	🌰	freestone
Harken	🌰	freestone
Madison	🌰	freestone
Red Haven	🌰 🌰	freestone
Heirloom		
Nectar	🌰	freestone
Red Haven	🌰 🌰	freestone
Polly White	🌰 🌰	cling

season. They especially need to be planted on sunny east-facing locations and protected from prevailing winds. Only a few selected microclimates at 3,000 feet and above may work in Zones 4–9. The flower bud is highly susceptible to May–June frosts. Nectarines (peaches without fuzz) are even less hardy. Look for varieties that have high chilling requirements (800–1,000 hours) in order to set flower, then fruit, because they will be the last to bloom after the danger of spring frosts.

Unlike apples, peaches need heavy pruning annually to produce well. Peaches bear on new wood (that year's growth), and the older years' wood should be pruned out. Because peaches are susceptible to lead-curl (aphid) on the new leaves of young trees, they must be sprayed with a lime-sulfur and dormant oil spray (purchased at your local nursery) when the flower buds first crack open in late November–December and then 3 weeks later; if that's not possible, then as soon as you can thereafter, in order to kill the eggs and larvae of the aphid.

Pears

European pears are the next hardy fruit tree to select. The Sierra above 3,000 feet are too cold for Asian pears. Pears are usually in Zones 4–9, so those of you below 4,000 feet will have better success at growing them. Pears need pairs to pollinate (male and female trees), and it's best to select a couple of varieties like a Bosc or an Ure help with pollination. Blooming begins in late April, so frost can be a problem. Pears are pretty much grown similarly to apples, and fruit around late September–October. See Table 6.4.

Plums

There are three types of plums: Japanese, European, and American. Japanese plums were imported from Japan and are the most productive and easy to grow. European and American plums are native to Europe and the United States respectively, and are harder to find and less productive. Plums in general bloom a good week or two before apples, and this makes them susceptible to spring killing frosts. Japanese and European plums grow well in Zones 5–9, while American plums are the hardiest to Zone 3. Plums, like cherries and apricots, like well-drained soils, and not wet clay soils. Apricots can be grafted to plum varieties. It's very important to pick plums when the flavors are sweet. Some varieties are tart and won't sweeten much so pick them when firm and not mushy. Store at room temperature short-term and in cold storage 38–41 degrees for long term. Table 6.5 shows some plum varieties to choose from.

Table 6.4 Suggested Pear Varieties			
Variety	*Region*	*Fruit*	*Uses*
Golden Spice	🌰	green russet	eat
Flemish Beauty	🌰	yellow russet	eat
Warren	🌰	green	eat
Dave's Delight	🌰	yellow russet	eat
Harrow Delight	🌰	green russet	eat
Lucious	🌰	yellow	eat
Parker	🌰	green russet	eat
Patton	🌰	green	eat
Heirloom			
Ayer's	🌰🌲	yellow russet	eat
Beurre Goubalt	🌰	green	eat
D'Anjou	🌰🌲	green/late	eat/cook
Dutches De Angouleme	🌰	yellow russet	eat
Humbug	🌰	green	eat
Malakoff Blush	🌰	green/small	eat/storage
Seckle	🌰	green/small	juice/cider
Blush Bartlett	🌰	green russet	eat
Beurre Gris	🌰	green	dessert/cook
Forest Asian Pear	🌰🌲	yellow/brown	eat/storage
Mother Load	🌰	yellow	eat
Organic			
Hardy (Beurre)	🌰🌲	yellow green	eat
Kieffer (Heirloom)	🌰🌲	yellow	eat
White Doyenne	🌰	yellow russet	eat

JULY & WINTER

Table 6.5 Suggested Plum Varieties		
Variety	*Region*	*Fruit*
Mount Royal	🌰	Japanese
Mirabelle	🌰	Japanese
Superior	🌰	Cross
Obilnaja	🌰	Cross
Sprite Delight	🌰	Cross
Alderman	🌰 🌲	American
Wantea	🌰 🌲	American
Heirloom		
Clairac Mammoth	🌰	French Prune
D' Ente	🌰	French Prune
Organic		
Oullins	🌰	European

PLANTING FRUIT TREES

Once you have selected your species (apple, pear, plum, peach, apricot, or cherry) and your varieties, go to the nursery and buy them. You can either order them in winter as bare root for the following spring, or you can buy them in pots later during the summer. It's cheaper to buy bare root and plant them in the spring, or if you are only looking to buy a couple trees, 5-gallon container trees can be planted anytime in the summer or fall. For bare root trees, it's best to select whips (small 3–4 foot) seedlings that are single-leader (unbranched), 2-year-old trees; they are the cheapest. The trunk should be straight and it should have a **caliper** (trunk width) of one-half to three-quarters of an inch in diameter. The tree should look healthy with no dead or broken leaders or tips. Before planting, soak them in a 5-gallon bucket of water for 12 to 24 hours to hydrate the roots, with a few tablespoons of B-1 fertilizer (liquid) mixed in as well.

The next step is to lay out your trees. 🌲 If you live in the Lake Tahoe Basin or in a heavy snow area, be careful not to plant your trees too close to the road or driveway where the blowing of snow from snow blowers (personal or city crew) could damage your trees.

Spacing depends on the type of tree that you purchased. Standard form trees are the largest, semi-dwarf next, and the small ones are dwarf. Most of your fruit trees will fall into these categories. For row spacing, in between trees: dwarf: 8 feet apart; semi-dwarf: 12 feet apart; and standard: 18 feet apart.

When planting, dig a hole about one and a half times the size of the roots, usually about 16 inches wide and about 18–24 inches deep. If you have a lot of trees to plant, rent an 18-inch hand auger, or if possible, a tractor.

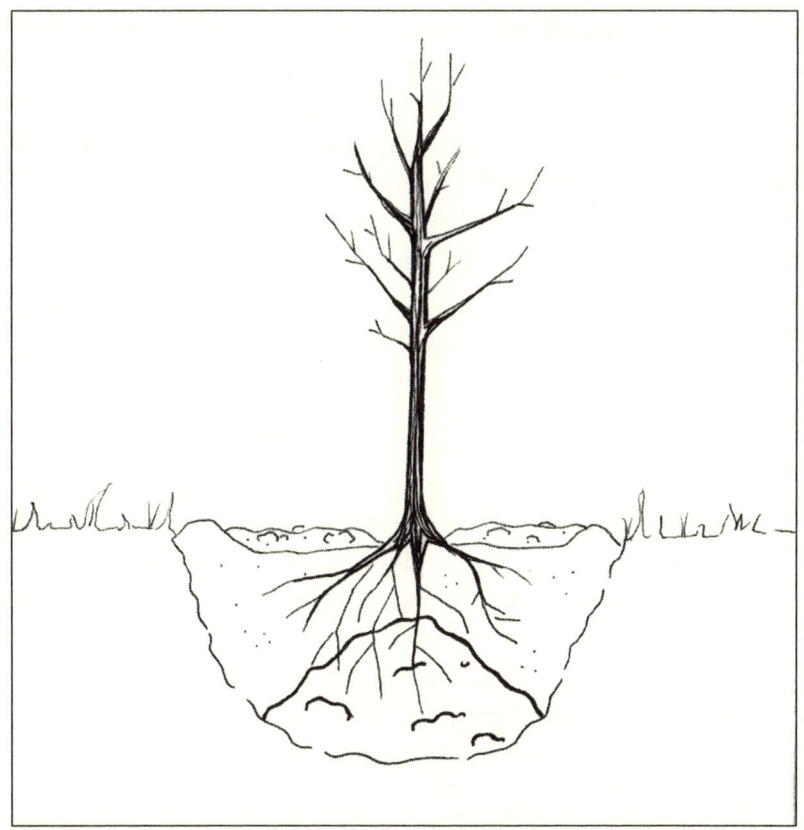

When planting, dig a hole about one and a half times the size of the roots. The crown of the tree should be slightly higher than ground level.

Add about 30 percent compost or soil amendment to the existing soil and about 6 ounces of bone meal and mix it all together. If the soil is somewhat dry it is best to fill the hole with water, let drain, then go ahead and plant the tree. For bare roots, place a small pile of soil at the bottom of the hole, spread the bare roots over it, and plant the tree. The crown of the tree (where the roots join the trunk) should be slightly higher than ground level when you finish back-filling the hole, and the graft union should be facing toward the prevailing winds.

Tamp the soil firmly around the tree with your foot and shovel a 3-foot-wide water basin around each tree. It's best to deep soak the trees at least two times immediately after planting to allow the soil to settle (each tree should receive about 6–8 gallons of water); after settling, top-dress the wells to bring them back to ground level, install your drip irrigation (one 2-gallon emitter on each side of the tree spaced about 1 foot away from the trunk), and mulch the tree well with pine needles or bark chips (composted), straw, or compost. Finally, mix up a whitewash to paint the trunks of your trees. The whitewash mix reflects the heat of the sun on the tender bark, preventing sunscald, which can crack the trunk of trees and allow entry for diseases and pests. Mix 3 parts water with 1 part interior latex paint and paint the bottom 24 inches of the trunk. This is best done annually in spring or fall.

Never give nitrogen fertilizer to newly planted trees. Allow them to come out of dormancy (bare root) and begin new leaves (which usually occurs when the soil reaches 45 degrees) before adding a simple low nitrogen fertilizer; a triple 10 is good (10% nitrogen-10% phosporous-10% potassium), or any of the organic formulations. The optimum growth for fruit trees is usually one month before bloom, and after the terminal buds set from July through August.

TREE CARE

Once the trees or whip (a single-leader sapling that is about 3–5 feet tall) are planted, just cut the top 6 inches off the top branch or single-leader tree, and 25 percent of the top of the tree in its second year. Always cut the branch about one-eighth of an inch above a bud, and select buds that face toward the outside of the tree and into the wind. If you are a novice, it's best to do less pruning than more. You can always come back later and do additional pruning. Go online and find pruning demonstrations, or attend nursery pruning classes. Participating in a master gardener program in your area is a great way to network with other gardeners and farmers and learn about tree care and pruning techniques.

Most fruit trees are pruned to a central leader system; you have a main trunk up through the center of the tree and lateral branches come off that. You want to balance a tree so that it is not lopsided, and you first want to remove any dead, dying, or broken branches (snow damage); from there remove any limbs that are growing inward on the tree. The objective of a healthy tree is to have an open center and wide branches with wide crotch attachments to the trunk and lateral limbs (narrow crotches can cause limb breakage during the winter months).

Most fruit trees described require about the same amount of water per season. The key is understanding your soil type, how much rainfall received per year; remember the west slope of the Sierra is wetter than the east slope,

Fruits, Nuts, and Berries

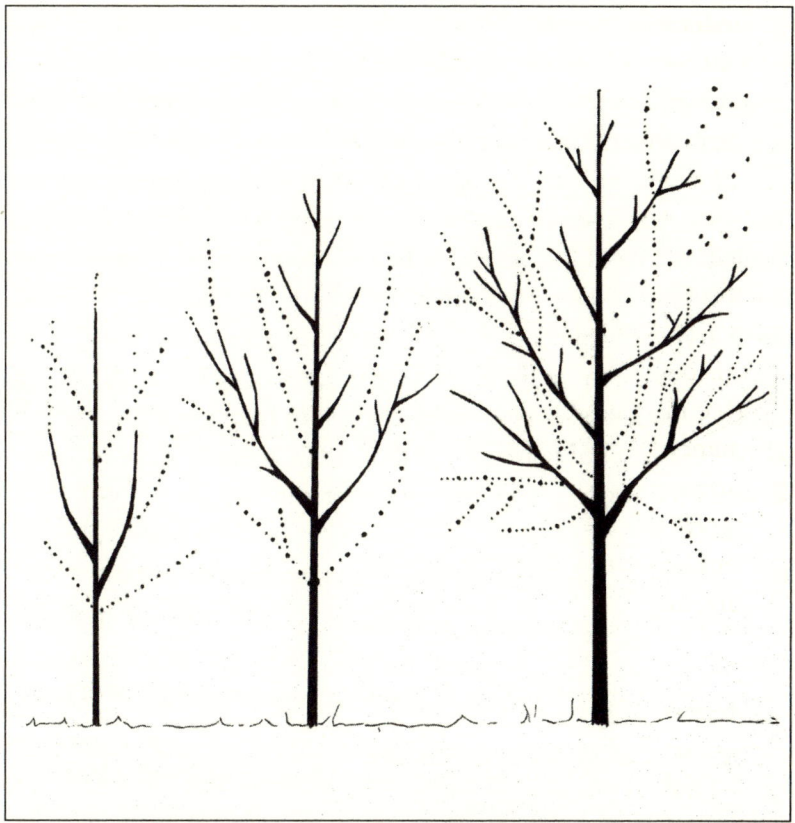

Pruning a tree at three stages of growth. Prune the tree to a central leader system so that it is not lopsided. Remove any dead, dying, or broken branches.

and the exposure and slope your trees are planted on can be a factor. In general, fruit trees require 20–30 inches of water per season. During winter, when the trees are dormant, it is not so much a factor, although during cold dry winters or late winters, early spring soaking may be needed if the soil becomes dry around the drip line of the tree (canopy width). In most cases, fruit trees require 2 to 3 inches of water per week during flowering and fruit set, and that same amount prior to harvest to maintain a plump juicy fruit.

There are many ways to irrigate; drip irrigation is best for small trees under 5 years old, then transfer to a microsprinkler that can spray 5–8 feet to water the drip line of the tree. Remember fruit trees have a fibrous root system, not a tap root like conifers, and their roots expand out to the canopy edges of the tree—it's more important to disperse water outward from the trunk, not just one or two emitters at the trunk from fruit-producing trees. Deep watering cycles are a must because roots follow water and the deeper the water goes the deeper the roots will follow. After deep watering it is best to let the soil dry out and then apply a deep soak after a couple of weeks. If you have sandy, gravely soils, you may have to water weekly during the hot summer months.

Once established after a couple of years, fruit trees are not heavy feeders of fertilizers, for conventional farmers and gardeners a triple 10 or 15 once in the spring is sufficient, and you can apply a 6-20-20 after harvest going into winter. For organics, apply a spring application of compost or organic blood meal, or feather meal, then end the season with a manure and bone meal application. Fall is a good time to disperse your fresh manures or compost around the fruit trees when going dormant, and allow them to decompose through winter and also provide insulation for the roots. Cover crops of peas, vetch, and mustards are great to plant between rows after harvest to help build your soils, reduce weeds, and help dry out your soils in the spring. It really helps to have a cover crop between the rows so you can have foot and vehicle access to prune and spray your trees without bogging down in the mud.

> Deep watering cycles are a must because roots follow water and the deeper the water goes the deeper the roots will follow.

With all fruit trees there is a spray element to your program, organic or not. The key is what you spray and timing. The insects you will likely encounter are aphids (leaf curl, peaches), codling moth (apples, pears), leaf rollers (all), spider mites (all), and apple maggot (apples), among others. The key is what you can live with. Dormant oils with an organic insecticide are best sprayed at least two to three times between November and bud set in the spring. An integrated pest program of spraying, trapping, and releasing beneficial insects is the best management practice for maintaining fruit tree orchards. The details are beyond my expertise, so check with your local ag advisors and UC Extension Office. Other pests can include birds, squirrels, deer, raccoons, and bear. (Please refer to Chapter 9 "Pests: The Good, the Bad, and the Ugly.")

There are a lot of repellents, but most work only once or twice, then the animals get smart and figure them out. Seclusion is the best long-term solution, using netting and fencing, or if you can contain livestock among your trees and orchard, you can use guard dogs to keep the critters away and kill two birds with one stone.

For weeds, it's best to keep a 4–5-foot weed-free ring away from the trunk and mulch it with composted bark, or even gravel (keeps the voles away from girdling, chewing the bark off the trunk of young trees). Creating a perennial cover crop between rows and mowing or grazing them is the best practice in the Sierra because perennial grasses and forbs are native to the area and establish by themselves. A pasture mix of orchard grass, crested wheatgrasses, perennial rye, blue grasses, bromes, and fescue can also be established to allow rotation grazing of cattle, sheep, hogs, and chickens to help with your nutrition program and allow chickens to clean up the insects after the ruminants have gone through. Be careful of using goats, because they like to climb trees and eat bark.

In the Sierra, the harsh environments can demolish fruit trees if precautions are not taken. 🌲 Winter damage to fruit trees can be a problem. If you are in a heavy snow-load area and have a few precious trees, take green staking tape and wrap the canopy into a conical shape to prevent limb breakage due to heavy snow. It is also a benefit to prune the top one-third of your annual growth after harvest before winter, so that you have a shorter leader that is more rigid and cannot be bent over and broken by heavy snow accumulations. Pruning your trees for an open center within the tree canopy will allow snow to fall through and not accumulate and cause freeze damage and limb breakage. In all cases, the trunks of fruit trees should be painted with a whitewash from below the first branching down to the ground to prevent sunscald, which is a freeze-thaw process that cracks the bark of fruit trees during the cold spring season and the hot-cold summer. This should be done every spring or fall. In addition, a trunk guard (purchased at nurseries in paper, cardboard, or plastic), along with a 5-foot weed-free zone around each tree, should be put on all fruit trees in the fall to prevent trunk girdling during the winter by voles, mice, and rabbits.

SPECIALTY FRUITS

From here on out, everything in this chapter is experimental, and you have to be in the right spot, a special microclimate within the Sierra, to grow any of these. 🌰 This group of specialty fruits are for gardeners or farmers in the 3,000-foot range. Let's get a few out of the way first. Forget citrus (lemon, orange, grapefruit), and even mandarins (which are the hardiest planted up to 1,200 feet, and can handle lows down to 20) cannot handle 3,000-foot elevation . . . sorry.

The next most sensitive would be your pomegranate, kumquats, persimmons, guava, and kiwis, which can only can handle heavy frosts down to 26 degrees below 1,000 feet in elevation. But here are some fun trees to play around with that can work in the right place.

Figs 🌰

Everyone wants a fig tree. Figs are the "in" thing right now. Well, those of you below 3,000 feet are in. Most of us are out. But there is hope. There is a lone fig tree along the middle fork of the Feather River, at about 2,800 feet, that I stop and pick wonderful figs from every September. It's a Black Mission, I believe. So here is my suggestion for figs. Select a north or western exposed site, by a water body (reservoir, pond, or stream). Mulch or place river rock around the base to absorb heat for the roots, and prune it small; wrap it into a conical shape with green tree wrap; cover it from late December through February, and hope for the best. The hardiest fig varieties to try are listed in Table 6.6.

Table 6.6 Suggested Fig Varieties		
Variety	*Region*	*Fruit*
Excel	🌰	black
Norland Fig	🌰 🌲	black (–10 degrees)
Heirloom		
French Paillard	🌰	purple/green
Ram Ranch Golden	🌰	white
Donna's Fig	🌰	hardy purple
White Adriatic	🌰	white
Organic		
Violette De Bourdeaux	🌰	purple

Chicago fig (*Ficus carica*) is an interesting one. It can freeze to the ground and be cut back to regrow and produce a crop in the same season. Best to cut back early winter and mulch root crown, then protect it in early spring.

Hardy Kiwi 🌰

This is a fun one; they usually don't persist too long in the Sierra, but may in the right spot. Hardy Kiwi (*Actinidia kolomikta*) is native to Russia, Zones 5–9. They prefer sunny locations out of the wind, are vigorous but vineless, with brown fruits ripening in late August to September. If picked hard, they will store for months in cold storage, and then can be brought out and left at room temperature to ripen in 7 to 14 days. Hardy Kiwi takes 3–5 years to bear fruit, and you need one male plant per every 8 female plants to pollinate the females. Like grapes, they need to be pruned heavy annually. Artic Beauty is the preferred variety.

Hawthorn

Hawthorns are related to pears and also have small berries and can produce wonderful flowers and berries. Very adaptive in the 5,000–6,000 foot range. Chinese Hawthorn (*Crataegus pinnatifida*) is a great choice for Zones 4–9.

Hops

Hops (*Humulus lupulus*) are a great crop for the Sierra. My grandmother had a hop plant growing on the side of the old bunkhouse for more than 40 years. They handle down to 30 below zero! We now have established an half acre hop garden of the Cascade variety here on the farm and it is doing very well! They are a vigorous vine that grows 10–12 feet in one season, produces a hop (fruit), which is dried and used in flavoring beer. They are much in demand and there are many varieties. Hops like full sun and protection from the wind. The root crowns should be covered with river rock to help protect them from winter damage. They are great grown on sides of buildings, or trellises to provide shade and cooling. They must be cut back at the end of harvest to the crown. Commercially they are grown on 15–20-foot pole and wire structures, and the hops are harvested in late September. Hops are very susceptible to spider mites in dusty regions. Hops are sold by rhizomes (underground portions of the root) that can be purchased online from hops.com, or from local nurseries. *Cascade, Nugget,* and *Chinook* are the best variety for the Sierra.

Mountain Ash (*Sorbus*)-Hawthorns (*Crateagus*)

Mountain Ash is a native to the Sierra and can be found through commercial nurseries. It has beautiful red berries that attract birds, especially woodpeckers. The fruits make make nice jellies. Found in Zones 3–8, a small, under-story tree. A variety, Shipova, is a cross between a pear and a mountain ash that is similar to a crabapple. Varieties to try are Ivan's Belle and Ivan's Beauty.

Prickly Pear Cactus

This is for you east-slope dwellers in the Sierra. *Opuntia cycloids* loves sandy, gravely, hot dry slopes. Found in Zones 6–10, it makes a beautiful reddish-pear fruit at the top of its beaver tail-like stems. Be careful! They are covered with thousands of hairlike spines and must be picked with heavy rubber gloves; then the spines must be rubbed off and rinsed thoroughly. The fruit then can be peeled to reveal the beautiful pulp within. They have a lot of seeds.

Quince

Quince is an under-utilized fruit crop and a dual-purpose shrub. The fruiting varieties (*Cydonia*) make beautiful quince (2–4 fruit per cluster) in late September and make great jellies and jams, and the flowering varieties (*Chaenomeles*) are great cut-flower crops for farmers markets and for selling wholesale to florists during the Chinese New Year in January. Great for Zones 5–9. Flowering quince needs two different varieties to pollinate; varieties come in red, pink, and white and will stay in bloom for more than a month. In addition, they are very thorny, make great windbreaks, and provide wonderful habitat for birds, especially quail.

NUTS AND SEEDS
Nuts

Again, let's get a few out of the way from the start. Once you get over 1,500-foot elevation you can pretty much forget about pecans, pistachios, hazelnuts, filberts, and peanuts. But you can consider these chosen few.

Acorns

Now here is a forgotten nut, the acorn. The mainstay of the American Indian. The acorn comes from the oak tree (*Quercus*), and there are many species found in the Sierra, usually below 4,500-foot elevation, mainly on the western slopes, in the mixed evergreen and conifer/woodland zones. The most valuable, because they contain the highest protein content of all the acorns, are the Black Oak (*Q. kelloggi*). Others that can be used are Canyon Live Oak (*Q. chrysolepis*), scrub oak (*Q. durata*), and Tanbark Oak (*Lithocarpus densiflora*). The acorn is collected in the fall, and must be gathered as soon as it begins to fall from the tree because it will quickly be consumed by deer and rodents and be infested by the oak moth larvae. Once collected, place the acorns in a 5-gallon bucket of water and leave overnight. The next morning the ones that floated to the top are not viable; the ones that sank are the ones you keep. You can then directly plant your acorns into containers.

Almonds

Believe it or not, there is a chance of growing almonds in the lowest part of the west slope for my region between 2,500 and 3,000 feet. Try Halls hardy variety from Trees of Antiquity, it is the hardiest of all the almonds, along with Texas Almond from the Felix Gillet Institute, which grows at 2,600-foot elevation. You just might be in the right microclimate to get away with it! Both are late bloomers and need 600–800 chill hours (the number of frost hours a fruit tree needs to be in dormancy before it leafs out in the spring in order to produce fruit). That's a good-sized winter!

Chestnuts

There are three kinds of chestnut trees (*Castanea*) that grow between 2,800 feet and about 3,500 feet, Zones 5–9, if placed in the right spot; the American (*Castanea dentata*), Collosal, and Nevada chestnut. The American chestnut is the most common. They are a large tree and prefer a mixed coniferous forest environment in canyons, small draws, and along streams. Two different grafted varieties must be planted to pollinate each other. They like acid, well-drained, sloped sites. They take 5–7 years to produce nuts, and they produce a lot of nuts. Table 6.7 lists a few varieties to try.

Table 6.7 Suggested Chestnut Varieties		
Variety	Region	Fruit
Nevada		American
American		American
Heirloom		
Donna De Lyon		European
Organic		
Collosal		American

Pine Nuts

The Sierra has an abundance of different kinds of pine trees throughout the mountain range, but only a few native species have valuable pine nuts. The native species are Gray Pine (Digger Pine, *Pinus sabiniana*) at 3,000 feet; the Sugar Pine (*Pinus lambertiana*) on the west slope and Single-Needle Pine (*Pinus monophylla*) on the east slope are found at the 5,000–7,000-foot range. All the pine nuts from these cones can be extracted in late fall, roasted and used in place of commercial pine nuts.

Pine trees can take 10–15 years to produce cones, so it's best to find established groves to pick from. It is illegal to collect cones from federal lands like the US forests and BLM (Bureau of Land Management) lands; a permit can be purchased from them, but it is very hard to do unless you are a commercial seed collector.

The nuts can be pried out of the cones with a knife or heated in an oven at 150 degrees to open the cones. Most can be put up in a warm attic for a couple of months and they will open on their own. The east slope of the Sierra is blessed with tons of single-needle piñon pines. The trees can be started from pine nuts, but they take forever to grow, and may take as much as 10-plus years to develop cones.

Walnuts

English walnuts (*Juglans*) will not make our list, but you may get away with some black walnuts. They are the rootstalk for the English walnut, the walnut you buy in the store. In commercial operations the English walnut (nut-producing part) of the tree is grafted to the black walnut (root stalk) because of its hardy root structure, vigor, and disease resistance. The black walnut is a large tree that will survive Zones 5–9. They produce a cherry-sized hard nut that is hard to crack, but has delicious meat inside. Table 6.8 provides a list of walnut varieties.

Table 6.8 Suggested Walnut Varieties		
Variety	Region	Fruit
Chambers #9		black
Maurogian		black
Heirloom		
Mega Leaf		English

Seeds

Quinoa

Quinoa can be grown in the Sierra at any elevation. At the higher elevation it's best direct-seeded around late June through early August. The limiting factor of quinoa is heat. It does not like temperatures over 90 degrees. If the quinoa plant is flowering in temperatures at 90 degrees or above it will go sterile and produce no seed. It's best grown on the west slope at cooler afternoon temperate areas. Once the seed is collected the saponin (seed coat) must be thoroughly washed and leached off the seed before using. Washing off the saponin is a difficult process for the home gardener because it takes a lot of time and if not done thoroughly can cause irritation to the throat when the seeds are eaten.

Sunflowers

Sunflowers do great in the Sierra in all zones, and all commercial food seed and ornamental sunflower varieties will flower and produce seeds. Just place a brown paper sack over the seed heads and tie them off when the seeds develop so that the songbirds don't get them before you do. The seeds can be roasted and salted for human consumption or added to your favorite birdseed and poultry feed mixes, and the ornamental varieties are great hosts for beneficial insects.

BERRIES

The Sierra offers a lot of opportunity for farmers and gardeners to raise all kinds of berries, especially below 4,500-foot elevation. Not only are they high-dollar crops, they also increase biodiversity, attracting wildlife and beneficial insects to your farm or gardens. The one issue with berries is that they are invasive, very labor intensive in pruning, harvest, and post-harvest, and they require refrigeration because they are extremely perishable. When it's berry season, you have to pick berries daily, and the season can go on all summer and into fall depending on the types and varieties you choose to grow.

Just like fruit trees, the limiting factor will be the June frosts at the higher elevations (5,000 feet and above). Most berries will survive the harsh winters of the Sierra, but the cold frosts of June can kill the flower buds, thus eliminating the chance for a fruit crop. Covering the berries with frost protection fabric or plastic during the danger of frost, or growing the vines in hoop houses are other alternatives. With all berries it's best to plant a number of different types and varieties so that you can have early, mid-summer, and fall harvests. If frost zaps one variety, another might just time it right and miss it. You have to try and catch that window of opportunity (weeks without frost), so that your flowers do not get damaged and you have fruit set.

In general, most berries like a loose, well-drained loam soil, somewhat acidic, and in a sunny location. Avoid wet, clay soils that when saturated can have standing water. Most berries are susceptible to root rot fungus like oak root fungus, phytophthora root rot, and verticillium root rot. In all cases, avoid planting any berries where tomatoes, potatoes, eggplant, okra, or peppers were grown in the same soil over the past 4 years—they are hosts for verticillium root rot.

For all berries, mid-summer watering is a must, and especially during berry development and harvest you want to apply 2–3 inches of water per week to keep the berries plump, full, and luscious. It's best to use drip irrigation, and place emitters at each berry plant. Overhead watering will cause fruit rot, mildew problems, and invite other bacterial diseases to your berries.

All berries must be picked almost daily; select only the ripe berries. Let berries ripen on the vine and pick wearing soft cotton gloves, as they will stain your fingers. They are best picked in the mornings; if the berries are picked in the afternoon they will retain the heat of the day and continue to ripen until cooled in a refrigerator. All berries should be refrigerated immediately at 33–36 degrees.

How to Grow Berries

For those of you above 4,500 feet, it's going to be a challenge to have consistent years of success with berries because of our unpredictable climate. Either way, one thing is for sure, you are going to have to compete with birds, squirrels, chipmunks, deer, raccoons, and bear for those berries, so be prepared to net most of your crops to prevent the feathered and four-legged thieves from stealing your crop!

Blueberries/Huckleberries

Blueberries and huckleberries are a very specialized crop that needs special soil and climatic conditions. Blueberries and huckleberries are a better choice for the northern regions of the Sierra; say around Shasta/Trinity, Lake Almanor and Chester/Westwood regions, where you have more moisture or lower elevations on the west slope of the Sierra below 3,500 feet. Huckleberries (*Vaccinum*) are native to the higher elevations of the Sierra, and are a valuable food source for songbirds, quail, and robins.

Blueberries are quite the investment; the plants are expensive and they take 5–7 years to reach full production, with no crop the first 2 years. Blueberries and huckleberries are acid-loving plants and prefer rich organic, loose soils with a pH range from 4.5 to 5.5, and they are quite sensitive to that range. Plants tend to be small, and the flower buds are very sensitive to late spring–early summer frosts. It may be wise to construct a simple temporary hoop house structure over your rows of blueberries and cover them with frost protection fabric during flowering to protect them from frost. Later you can use bird netting to protect the berries from birds.

When planting blueberries or huckleberries, space plants 3–4 feet apart down the row with 8–10 feet between rows. Add about 30 percent peat moss to the planting hole to ensure the acidity, then water and mulch with native forest litter (pine duff, needles, and so on). Most cultivars are self-pollinating. Annual pruning is required to cut out dead or damaged branches and spindly twigs. Prune lower branches up about 6 inches from the ground, so that fruit does not touch the soil. They are pretty much insect free; the biggest problems in the Sierra are birds, rodents (squirrels, chipmunks, voles, rabbits), raccoons, deer, and bear. Especially bear—they will strip all your berries in a few nights. To protect them, use seclusion fencing, invest in a guard dog or two, or place some livestock around the perimeter of your blueberries/huckleberries when they are ready to harvest.

There are a number of varieties of blueberries, but the best selection for the Sierra is the Minnesota low bush varieties that fruit about mid-summer. These half-high varieties, as they are known, have been bred to withstand heavy snow loads and colder winters than other varieties and are hardy for Zones 3–7. These varieties are North Country, North Sky, North Blue, Northland, Bonus, Elliot, and Patriot from Indiana Berry Company.

Huckleberries are larger plants than blueberries, require very little maintenance, and make a great hedgerow around your blueberries for wind protection. Huckleberries are usually sold as Evergreen huckleberry (*Vaccinum ovatum*) at local nurseries in your area.

Cranberries Lingonberries

The American cranberry (*Vaccinum macrocarpon*) is another acid-loving plant that may have some potential in wetter sites around the Sierra. Cranberries don't have to be associated with bogs, but they do need an abundance of water, and they like to keep their feet wet. Hardy to Zones 2–7, cranberries can withstand –45 degrees. Cranberries propagate by sending out 3-foot runners in which the plantlets root when in contact with the soil and create a matlike, 1-foot groundcover. Given the right spot, cranberries may play a part in your garden or on your farm.

The Lingonberry is kind of an odd one, but interesting. It is related to cranberries and blueberries and is an evergreen small shrub. It's another acid-loving plant that produces a slightly tart red berry that's harvested in the fall. I haven't tried this one, but it may bloom after the last frost at elevations between 3,000 and 3,500 feet in the right spot. Lingonberries need well-drained soils and very little fertilization. The berries are used for jams, jellies, pies, and pastries. They prefer full sun and are about the size of blueberries, but a rhizome sends out shoots that create an 18-inch tall ground cover or sprawling shrub. The interesting part is that they can be cut back for winter and mulched, then resprouted in spring and produce a crop for fall. Hardy to Zones 3–9.

Brambles: Raspberries and Blackberries

Called brambles, raspberries and blackberries are in the same genus *Rubus*, which are perennial vines that produce fruit on annual vegetative canes (*primo canes*), and second-year woody canes (*fornicanes*), which then die after the second year and must be pruned out. Brambles need heavy pruning annually to keep the vigor and fruit productivity of the plant for the next year. Most brambles are productive for more than 10 years, and it's best to remove any of the wild or Himalayan berries within 600 feet of your cultivated varieties of brambles to prevent diseases from entering your crop. Brambles like well-drained soils, slightly sloped, and grow best in a pH of 5.8–7.0.

Most brambles are purchased from nurseries as bare root plants or in 1-gallon pots in early spring and should be planted 2 inches deeper than the crown so that the buds will grow from underground versus above ground. Two to three buds are left above ground, and the rest of the cane is cut off and watered immediately. Brambles don't need to be trellised the first year and don't need a lot of fertilizer, so a spring and fall application of compost or composted manures works well. Bramble varieties have thorns or are bred to be thornless. The thorny varieties tend to be better suited for our cold climates, and raspberries (Zones 2–7) are better adapted to the Sierra higher elevations above 4,000 feet than are blackberries (Zones 4–9).

Blackberries differ from raspberries in that they are a little less hardy and come in varieties of thorned and thornless. They can be pruned as a bush or trained on a trellis. They too have to have all their floricanes cut back to ground level at the end of the year and burned or mulched. In the spring primo canes can be thinned to 5–6 canes per lineal foot of row if grown erect; if attached to trellis leave 6–8 canes. Blackberry varieties include Illini, Prime Ark, and Chester, Ebony King, all from Indian Berry Company.

Raspberries can be found in red, gold, purple, and black. Most varieties are red. Raspberries are not quite as aggressive as blackberries and can be planted 3–4 feet apart along a row with 10–12 feet between rows; they are a better selection for higher elevations. They can be grown erect like a bush, but do best if trained (for ease of picking) on a T-shaped trellis (24-inches wide), with a strand of wire or twine on each side of the T, strung about 5 foot high parallel to the row of raspberries. For fall-bearing raspberries, each late spring to summer as the canes get 4–5 feet tall, they should be attached to the trellis and the top 6 inches cut back to a bud. After 3 years and each year thereafter, all the floricanes (2-year woody canes that bear fruit) should be thinned to 3–5 canes per lineal foot of row for the season. After harvest, all the floricanes must be removed (cut just below ground), and burned or mulched. For floricane production, black and purple raspberries can be pruned as a bush similarly to roses, leaving 4–5 canes and pruning back the lateral branches to 12 inches from the main floricane, thinned and trained on a trellis similarly to fall harvest raspberries. Table 6.9 lists your best raspberry varieties.

Currants and Gooseberries

Currants and gooseberries are in the Genus *Ribes*, and closely related. The difference is gooseberries (wild) have thorns and currants don't in nature;

Fruits, Nuts, and Berries

Raspberries are easy to pick if trained to grow on a trellis.

Table 6.9 Suggested Raspberry Varieties		
Variety	*Region*	*Fruit*
Boyne	🌰🍍	red
Canby	🌰	red
Killarney	🌰	red
Chilcoten	🌰	red
Nova	🌰	red
Autumn Bliss	🌰	red
Tulameen	🌰	red
Polara	🌰	red
Heirloom		
Heritage	🌰	red

there are hybrid thornless varieties of gooseberries commercially. There are many native species of currants and gooseberries in the higher elevations of the Sierra (see page 104); in this section I concentrate only on the commercial varieties. Currant juice is a major commodity in Europe. Most Europeans drink currant juice like we drink orange juice. The reason it hasn't taken on in the United States is because it was banned in the 1960s because it was believed to be a host for the white pine blister rust that was killing the western white pines in the Sierra. Since then it has been proven that currants don't contribute to the disease, but currants haven't found their way into the mainstream of the United States. This could be your niche market!

Currants are small berries about the size of a pea that ripen in late summer. They are used in syrups, jellies, pies, pastries, juices, and wine. Currants and gooseberries are small to midsized shrubs that grow in a variety of well-drained soils in a pH range between 6 and 7. They are hardy to Zones 2–6 and prefer full sun or filtered sunny conditions. Table 6.10 lists some currants for you to try.

Table 6.10 Suggested Currant Varieties		
Variety	*Region*	*Fruit*
Ben Serek	🌰	black
Black September	🌰	black
Black Down	🌰	black
Perfection	🌰	red
White Imperial	🌰	white
Organic		
Crandall	🌰	black
Primus White	🌰	white
Red Lake	🌰 🌲	red
Rovada Red	🌰	red

Most gooseberries in the wild have a spiny-fleshy coating around the berry that has to be removed and makes for a time-consuming process, but they tend to be a little sweeter than currants. Commercial producers have bred out the thorns, creating "thornless" varieties in both red and purple/black berries. Both currants and gooseberries can be planted 4–5 feet apart along a row, and about 8 feet between rows. The first year, cut the stems back to the ground and leaves 2–3 buds above ground. After the second year you can just remove dead or damaged stems, remove any spindly twigs, and thin the basal stems to 6 or 8 upright, sturdy stems, and yearly remove stems that are older than 3 years old. Gooseberries are labor intensive to harvest and are very perishable. They must be kept in cold storage. Table 6.11 lists some gooseberry varieties to try.

Table 6.11 Suggested Gooseberry Varieties		
Variety	*Region*	*Fruit*
Jahn's Prairie	🌰	red
Organic		
Pixwell	🌰	red
Hinnomaki	🌰 🌲	red and yellow
Black Velvet	🌰	red
Poorman	🌰	red

Grapes 🌰 🌲

You have table grapes and wine grapes. Everyone today wants a vineyard. I know I do, but fat chance here in Sierra Valley. I know University of Reno's agricultural experiment station has a wonderful research vineyard of all kinds of varieties. I'm not up on all the work they are doing, but they can probably get you a listing of the grape varieties best suited for the Reno/Carson City Area.

Reno is at 4,500 feet in the Great Basin, but merges 20 miles away with the eastern slope of the Sierra. It seems like it might be possible to grow red wine grapes in the southern portion of the east slope of the Sierra. I do know that table grapes are grown in the Lake Almanor/Chester and Greenville/Indian Valley portion of the Sierra, around 4,500–5,500-foot elevations on eastern- and southern-exposed sunny sites. When planted they must be placed 8 feet apart and have 10 feet between rows. All grape plants are usually 2 years old when purchased from nurseries and should be lightly root pruned and planted with the crown (where the canes meet the root stock) 1 inch below the surface of the soil. All canes should be cut back, leaving only two or three buds above the ground. Grapes must be trellised and protected from wind exposure and June frosts. It's best to plant multiple varieties for pollination. Prune all grapes when dormant during the winter months; prune out all older fruiting wood, and cut back laterals. Mulch the root crowns in the fall to protect them from winter freeze damage. In the spring, keep the four best-looking canes and prune out the rest; train two canes along the trellis in one direction and two in the other direction, leaving four to six buds on each cane. These canes then become the limb structure that you will cut the laterals back to in the fall. The following year you will select four more canes from this structure to carry the fruit for that year. Some older varieties to try at or below 3,000 feet are shown in Table 6.12.

Table 6.12 Suggested Grape Varieties		
Variety	*Region*	*Fruit*
Concord	🌰	table/wine
Muscat	🌰	wine
Niagra	🌰	wine
Catawba	🌰	wine
Buffalo	🌰	wine
Bath	🌰	wine
Canadice	🌰	wine
Glenore	🌰	wine
Himrod	🌰	seedless
Lakemont	🌰	seedless
Stueben	🌰	table/wine
White Diamond	🌰	white wine
Heirloom		
Lowell Hill	🌰 (3800 ft)	wine/vinegar
Squire Canyon Road	🌰 (3100 ft)	wine/red

Logan and Boysenberry 🌰

Both originated in California in the 1920s and are hardy to −10 degrees. Classified for Zones 5–9, they can be grown at the lower elevation of 3,000–4,000. Both are grown, pruned, and trellised similar to the other brambles. The best variety of boysenberry that adapts to all of the Sierra zones is *Rubus ursinus var. loganbaccus* from Trees of Antiquity.

Strawberries

Strawberries do very well in the Sierra, and there are a few native wood strawberries found up to 7,000-foot elevation. They are mostly in somewhat shaded areas, are smaller in size, and don't produce half as well as the cultivars, but they are sweeter. All strawberries will begin to grow in spring once the soil temperature reaches 40 degrees because they are shallow-rooted evergreens. In production, they are heavy feeders; so a fertilizer program of fall manures and a spring/summer side dressing of a balanced nitrogen, phosphorous, and potassium fertilizer is essential.

There are two types of commercially available strawberries: *June-bearing*, and *ever-bearing*, sometimes called day-neutral; though they are slightly different, ever-bearing and day-neutral are considered the same. Ever-bearing strawberries have the ability to flower and fruit in both short and long days, which allows you to get a crop the first year. A crop is planted in the spring, and the flowers are removed for the first 6–8 weeks to promote root growth, then strawberries will be harvested during the warmer parts of the summer, or fall; the plants will slow down for the winter, and then appear the next spring ready to fruit again. Ever-bearing strawberries usually last two to three seasons and don't need to be renovated and thinned like June-bearing ones, which are replaced. These strawberries reproduce by sending out runners that root to the soil surface that then can be cut off and transplanted into pots to create new plants for the upcoming year. All ever-bearing strawberries should be planted in raised beds with 8–10-inch spacing. On the second year I interplant the winter-rooted runners in the spring to a density of a strawberry plant every 6 inches. All beds should be watered with drip irrigation and mulched with straw or pine needles to create bedding for the strawberry. Ever-bearing strawberry varieties are better suited for our higher elevations and are cold hardy. Ever-bearing varieties grow well in Zones 4–9, and your best choices are Albion, Seascape, Tribute, Quinalt, and Tristar.

June-bearing strawberries are pretty much perennial season strawberries that will persist in your beds for up to 7 years. They do require maintenance to renovate, and thin your beds every 2 or 3 years. The crowns are planted in the spring and allowed to grow throughout the first season without a harvest. The flowers are pruned at the beginning of the year to develop vigorous plants and stimulate runners, which are pushed into the soil to root new plants and create a density of a plant every 6 inches until August. After August all runners and flowers should be removed until late fall. Before winter the strawberries are covered with straw or pine needles until spring, when they will begin to flower and set fruit. Keep a bedding (straw or pine needles) under the strawberry to protect the strawberry from coming in contact with the soil, which can cause fruit rot problems. The flower buds of June-bearing varieties are very susceptible to spring frosts, which can wipe out your crop outdoors. That's why they are ill advised for the higher elevations of the Sierra. One could place them in low hoop houses or cold frames for frost protection or consider them for greenhouse production to offset the season with ever-bearing varieties outdoors. The best June-bearing varieties are Cavendish, Annapolis, Itasca and Mesabi in Zones 3–8, with Shuksan and Honeoye the hardiest.

> Native berries don't produce as much as domestic varieties but they are more adaptive, require less maintenance, the fruit is smaller, and they have unique tastes and more nutrition than commercial berries.

Wild Plants and Berries

This subject has always fascinated me. As a native plant nurseryman for more than 20 years, propagating native Sierra wild plants for nurseries, highway projects, and reforestation and mitigation projects, I always thought that instead of trying to get cultivars to adapt to the harsh environments of the Sierra, why don't we just try and tame the wild berries and create a niche market for them? Native berries don't produce as much as domestic varieties but they are more adaptive, require less maintenance, the fruit is smaller, and they have unique tastes and more nutrition than commercial berries. If I was 20 years younger, I would set up a wild berry farm around the 3-4,000 foot elevation range! I've put a list of berries together for you, a wild edible list that you can experiment with in your gardens and around your farm. The key factor in all of these wild berries is to fertilize them with a balanced fertilizer; prune them as if they were cultivars; and give them plenty of water during fruit set and harvest. See what you can do from my list.

California Blackberry

Rubus vitifolius is found below 4,000 feet on the western slope of mixed evergreen and ponderosa pine forest along stream banks and wet areas. Berries are small but can be grown like commercial varieties; very thorny, but not as aggressive as the commercial varieties. Fruit is very sweet, somewhat dry.

Elderberry

The Elderberry (*Sambucus mexicana*, *S. caerulea* is a favorite of most people. A basal sprouting perennial shrub that dies back every year and sends out incredible canes annually in an irregular shape, with large palm-shaped leaves. The hollow canes were dried by the Indians and used for flutes. All parts of the elderberry are poisonous except for the flowers and the berries. The flowers can be dipped in flour and deep-fried, and the dark purple berries make great pies and preserves along with syrups. My grandmother made elderberry wine with the fermented berries, and it is still done today. Elderberries like well-drained soils, partial shade, and are found along stream corridors. Easily established in your gardens and around ponds, reservoirs, or wetter areas around the farm, they will attract pollinators, and beneficial insects and birds, along with bear (be careful).

Sand Cherry and Chokecherry

The sand cherry (*Prunus bessyi*; non-native) is an east slope cousin to the southern Sierra chokecherry (*Prunus virginiana*). A medium-sized shrub 5–7 feet tall, it prefers dry, sandy, gravely south-facing slopes and produces a drupe-like berry with a large pit. Good potential for adding some water and fertilization to create a berry for preserves, jellies, pies, and wine.

The chokecherry is a Sierra Valley favorite. A large black-colored berry, the size of a marble if watered and in partial shade, this 8-foot deciduous shrub can become an under-story tree in dense forests within the Sierra and makes a nice hedgerow. My grandmother made chokecherry pies and jams and jellies, also wine, and with proper fertilization, pruning, and ample amounts of water, growing chokecherries is very promising. The berries are very popular with birds and bears, and are a browse species for deer. In addition, they are great for pollinators and hosting beneficial insects, bees, and wasps.

Serviceberries

Amelanchier alnifolia, A. pallida grow in association with chokecherries and are common on the east slope of the Sierra and the higher elevations of the west slope around Lake Tahoe. A slightly smaller shrub than the chokecherry, it has a smaller purple/black berry about the size of a pea, dry in texture, that would make great preserves, syrups, juices, and pies if the shrubs were watered and fertilized regularly. A browse species for deer, they are a great border plant around your gardens for birds, beneficial insects, and pollinators like native bees and flies.

Sierra Plum

Yes, there is a wild plum tree *(Prunus subcordata)*, a small fruiting tree that is sparse across the Sierra. The Sierra plum produces a small edible red-purple fruit found below 6,000 feet between Sierra and Shasta counties and in the southern portion of the Sierra above Kern County. The key is finding the fruit and planting the pits. With some cultivation this could be a gem if the fruit can be protected from squirrels, birds, and bear.

Thimbleberry
Rubus parviflorus is a great native understory, shade-loving ground cover. It's found underneath many coniferous forests, up to 8,000-foot elevation and along stream banks. This papery, large palm-leaved perennial produces a thimble-like raspberry that is delicious. The problem is the critters like it too. If it could be raised in a protected environment, or netted, this could be a new crop for shaded areas. Propagates easily from rhizomes and annual new canes. Creates a solid mat-ground cover about 18 inches to 24 inches high.

Western Raspberry
Rubus leucodermis is a small vine, with purple-to-black small berries, found along stream banks below 7,000-foot elevation. It has a very fragile vine, spindly and not very aggressive, and does not produce a lot of fruit. It would be fun to experiment to see how well it does under cultivation.

Wild Currants and Gooseberries
The most promising of all the native berries. Common wild currants that you see in the Sierra are wax currant (*Ribes cereum*) at dry sites, Sierra currant (*R. sanguineum*, *R. nevadense*) in the shade, and understory and golden currant (*R. aureum*) on the dry east slope. Grow and manage similar to commercial varieties. All currants are pink flowering with purple berries except the golden currant, which is the first flowering currant in the spring with yellow flowers producing beautiful red/orange berries in September. Gooseberries and currants are great borders for insectaries and hosting beneficial insects, bees, and pollinators. The wild gooseberries are very spiny. Pick with gloves and peel open to get to the pulp. Sierra species is *Ribes roezeli*.

Wild Grape
Vitis californica is found along stream banks below 4,000-foot elevations on the west slope of the Sierra in the mixed coniferous forest. The vine produces a small cluster of purple fruits. It would be fun to experiment using this as a hardy root stock to survive the Sierra climate and graft on other varieties.

Wild Strawberry
Fragaria californica or *nigra* are well adapted to the Sierra and can be perennial crops within our gardens and farms. The fruits are smaller and sweeter, and the plants are less productive than commercial varieties, but they persist year after year. Small, protected raised beds can be a great addition as a groundcover within other berry crops.

Woods Rose

An interesting choice is the Sierra rose (*Rosa woodsii, R. gymnocarpa , R. californica*). I selected the Sierra rose for two reasons: first, its pink, edible single flowers, and second, for its fall rose hips. Rose hips contain 10–12 times the amount of vitamin C as an orange! We must be able to do something with the rose hip, like dry it for teas and make rose hip jam. It's a great thicket rose for your High Sierra gardens, and an erosion control plant for stabilizing slopes and stream banks. Very drought tolerant, it provides great cover for quail, rabbits, and songbirds and is browsed by deer. It's a great border for beneficial insects and pollinators.

Winter at Sierra Valley Farms.

CHAPTER 7
PUT ANOTHER LOG ON THE FIRE: EXTENDING YOUR SEASON

In any attempt to farm or garden in the Sierra, the consideration of extending your season is a must. Because of our short season, from 30 to 80 days at elevations above 4,000 feet, trying to grow any crops other than cool-season crops from June through September is difficult without some kind of season extension structure such as cold frames, hoop houses, or greenhouses. Unless specifically labeled as a greenhouse or cold frame I use the term "hoop house" as the basic nonheated structure you're using in the home garden or on a small farm.

The structure, or hoop house, can be as simple as adding frost protection fabric over a simple PVC frame to an elaborate commercial kit.

> When considering a hoop house, the first question to ask is "Do I want to want to grow year-round or just add a few months to the beginning and end of my growing season?"

When considering a hoop house, the first question to ask is "Do I want to want to grow year-round or just add a few months to the beginning and end of my growing season?" Most gardeners and small growers often just want a small hoop house or greenhouse to start their transplants in the early spring to get a head start on their tomato plants or other summer vegetables and herbs, or flower transplants. Another important factor in deciding what type of structure to use is to know what vegetable crops you can grow outside and don't have to protect. In areas below 4,000 feet in the Sierra, most residents can grow corn, tomatoes, squash, eggplant, beans, and other summer crops outside after the last frost in late May, so they may need only simple, temporary hoop houses that will protect them until those late frosts, or use them to get an earlier start on the season. On the other extreme, if you live at 5,000 feet or above, where during the winter you may have consecutive days of single-digit temperatures or days well below zero and many feet of snow, you will need a structurally strong greenhouse or geodesic dome to grow vegetables during the long winter months. In the winter the greenhouse or hoophouse structure must be sealed tight and able to handle high winds and heavy snow loads, and in the summer it must be able to be opened and closed to handle a 30-degree morning and a 90-degree afternoon. Or consider a geodesic dome.

The structure, or hoop house, can be as simple as adding frost protection fabric over a PVC frame to an elaborate commercial kit. This is a quonset-style hoop house at Sierra Valley Farms.

Site selection is the first component of creating growing environments within these structures.

SITE SELECTION

There are two theories on site selection of season extension structures. The ideal concept is to place the structure over the best piece of ground that you have, and the other is to place it over the worst piece of ground that you have and build the soils within the hoop house. It's a lot easier to build soils in a structure rather than outdoors in our cold climate because we can manipulate and control the environment within a hoop house. By extending the season and keeping warmer temperatures in the hoop house, you can take a hoop house out of production for 2–3 months in the summer or winter, add a cover crop, humus, composts, manures, and inputs of sulfur, limestone, or gypsum and build your soils over a short period of time. That is what we have chosen to do here at Sierra Valley Farms, so that we can save our good growing ground for outdoor crops. It's even possible to add a chicken or rabbit operation into a hoop house for a season, then turn all the manures under for winter and begin a great vegetable production program for the next season.

Here at Sierra Valley Farms I have five different greenhouse–hoop house combinations: a heated A-frame greenhouse covered in dual-pane Plexiglas for propagating and growing vegetable and native plant starts and microgreens year-round; a Quonset-style cold-frame hoop house on a gravel floor for nursery production and over-wintering native seedlings; a Quonset shade area to shade my native plants in the summer; a Quonset walk-in tunnel (caterpillars, as they are called), for vegetable production; a gothic-style, high tunnel that I can drive equipment through for farming vegetables; and outdoor raised beds with low tunnels used for perennial crop production of garlic, chives, shallots, and leeks. Each one has a purpose that enables me to farm year-round in one of the harshest climates in the West. Read on to select the type of structure that is right for you.

SEASON EXTENSION STRUCTURES

I cover most of the common structures applicable to the home gardener and small farmer. Most of us farmers have financial constraints that allow us to put together hoop house structures only with materials found around the farm under "ranch construction" (rough carpentry); if you are really serious about growing year-round I suggest *The Winter Harvest Handbook* by Eliot Coleman, the guru of growing year-round.

Cold Frames

Cold frames are simple window frame structures that were built in Europe decades ago to capture the soil's temperature to propagate seedlings in the spring and house root and stem cuttings of fruit trees and roses. They are the easiest to construct using the old Dutch-style glass cold frames constructed with old wood and hinged glass tops. The standard design of a bottomless box made of a 2 × 12-inch back wall (or rough 4-inch logs) and 2 × 8-inch front wall covered with hinged (30 × 60) glass panels is still the best. I like to situate a few boxes opposite of each other, a few facing east or southeast in order to collect the maximum amount of early spring sunlight for spring growing, and place another few facing north to northeast to cool my lettuces and cool crops for later summer growth. You can plant them with the same winter crops like lettuces or spinach and get a 1- or 2-week succession of harvest because of their different exposures to sunlight.

An easy-to-make cold frame.

You can enhance the cold frame boxes by framing them with cinder blocks, painting them black, and filling them with dirt to absorb more heat, or line the inside of the box with bricks. Cold frames will be safe down to about 24 degrees. If you have power available, you can add a couple of cross-bars inside the box and suspend a clear span of Christmas lights on a timer, or a piece of floating row cover, to give you a few degrees more. Cold frames are also great for hardening off your vegetable transplants. During the summer, if you can keep the windows fully open, the lower ground temperature will act as a cooling agent and you can extend your cool-season crops with micro-misters and some 50 percent shade cloth.

If you place one in your greenhouse or hoop house, cold frames are great for starting transplants of lettuces, herbs, and vegetable starts, or creating sets of chives, green onions, shallots, leeks, or garlic during late winter and early spring.

Hoop Houses

There are many types of hoop houses, or tunnels, as they are sometimes called, and I discuss two: low-tunnel hoop houses are those that you cannot stand in, and high-tunnel hoop houses are those that you can stand in, or even drive equipment through. By definition, a hoop house or tunnel are solar greenhouses that have no supplemental heat; they merely trap the sun's energy and maximize the Earth's insulation for the benefits of plant growth. A hoop house's only source of heat is the sun, and its only cooling source is manual venting, misting, or shading with shade cloth. Hoop houses were manufactured to extend the season primarily for vegetable production in the Midwest and eastern states because of their cold winters.

Low Hoop Houses

Low hoop houses are cheap, easy to assemble, and work well in extending your seasons indoors and outdoors. They are usually placed outside over a constructed raised bed, or indoors over individual crops for frost protection. In outdoor operations, a constructed raised bed should be no wider than 4 to 5 feet, so that you have easy access from either side of the bed, and 18 inches to 24 inches high. 🌰 In the Sierra, because of our high winds and heavy snows, I strongly recommend that you use half-inch electrical (EMT) conduit in 10-foot lengths as your hoops, and not PVC. There are two problems with PVC: (1) It will not stand up to the wind and snow loads, and (2) it's been proven that PVC chemically reacts with the UV-rated poly films and causes a breakdown in the plastics that will reduce the life of your film. Conduit is pretty cheap, and if you get a pipe bender from Johnny's Seed Catalog (www.johnnyseeds.com), the arc is already created for you. Place the hoops about 5 feet apart along the length of your beds, and just push them into the ground about 6 inches—secure them to your wood borders with U-clamps. In the winter, spring, and fall you can suspend floating row covers over them, and in the summer suspend shade cloth for cooler season crops or just leave them open for summer vegetable production. Once the fabric is placed over the hoops use plastic snap clamps to secure the fabric to the hoops, and in high-wind areas use sand bags or rocks to weight down the ends. If you sew a folded sleeve along each end and slide in a piece of PVC pipe, old broom, or farm tool handle, it will be easier to open the sides for harvest.

🌰 For winter vegetable production in the Sierra above 5,000-foot elevation, low tunnels in a greenhouse or inside a larger hoop house during the winter

A few Sierra Valley Farm season extension structures. A-frame greenhouse (top); Quonset-style cold-frame hoop house (middle); and Gothic-style high tunnel hoop house (bottom).

and early spring are a must. Use a heavy 10-gauge wire hoop, space hoops 4–5 feet apart, and clip the fabric to them with clothes pins. If you are growing lettuces or cool-season crops when the outside temperatures consistently get below 24 degrees, you want to create low tunnels over your crops inside the hoop houses and place 30 or 50 percent bond floating row covers over them. If it gets below 15, you want to add a full blanket of 50 percent over all of your crops and remember to open up the covers mid-morning (10:00 AM), and cover again at sunset.

High Hoop Houses

High hoop houses (high tunnels) are built so that you can walk inside them; some are even large enough to drive equipment through. In the Sierra, there are many uses for high tunnels; the obvious is to extend your growing seasons into the fall and get an early start in the spring. There are other uses as well. For the home gardener or small farmer you can build one yourself for about $400–$500 in material, a simple 8-foot-wide × 30-foot-long hoop house. Here's how you do it:

1. It's best to situate the structure broadside (directly sideways to the wind), so determine your dominant wind direction and place the hoop house door entrance end facing away from the wind. Your end walls are usually your weakest points. The width should be wide enough for a 3-foot bed on either side the length of the hoop house, and a 2-foot walkway down the center.
2. Construct hoops out of half-inch EMT conduit with 5-foot sides, and a 7-foot arched center, with seven hoops total, one placed every 6 feet, with a center purlin (reinforcing braces) down the middle of the arch, the length of the hoop house, wired to each purlin.
3. Each hoop upright is pushed into the ground 10–12 inches, and the corner hoops are concreted in.
4. Run a 2 × 6-inch redwood baseboard along all sides of the tunnel, and attach each hoop to the baseboard.
5. Construct a solid wall windward using either plywood or dual-pane Plexiglass, or polycarbonate with upper ventilation (vents), and construct a front door on the leeward end with upper ventilation (vents).
6. Cover the hoop house with a commerical grade 6 mil UV-rated film. To attach the film use a guide system from Farm Tek, a guide and wiggle-wire set-up, that is easy to put on and take off the plastic at any time of the year, or add shade cloth in the summer. For winter vegetable production you want to be able to completely seal out the

cold elements of the Sierra; the film should be measured to extend at least 12–18 inches wider than the width of the tunnel so that you can dig an 8–10-inch trench along each side of the tunnel and bury the film in the ground. This will give you winter insulation from the cold and discourage rodents from getting in.

The key to any hoop house or tunnel is ventilation in the summer and insulation in the winter and off-season. There are many different solar vents and fans available, and even adding a nice screen door for summer applications is a great asset.

🌲 Walk-in, 7-foot tunnels are great for the Sierra because they are low enough to handle the wind and snow, they are portable and can be moved, and they are versatile enough to use as a summer shade house. You can buy a number of prefabricated high-tunnel/hoop house structure kits from manufacturers. I strongly recommend that those of you gardening above 5,000 feet in heavy wind areas on the east slope around Carson City/Gardnerville, or in a heavy snow-load area around Tahoe or the Mammoth Lakes area consider an A-frame polycarbonate structure, Gothic-style, or a geodesic dome to handle Mother Sierra's inclement weather.

When considering the type of hoop house for the majority of gardeners in the Sierra, I suggest you select gothic-style houses, because they have peaked roofs that shed the snow well, and straight sidewalls that when snow is shed from the roof create an igloo-like effect that insulates the sidewalls. The walls of snow melt against the plastic during the day, and freeze at night, creating an ice wall that protects the plastic from the weight of the snow. The older Quonset-style hoop houses are not recommended for heavy snow areas, because they will collect wet snow and not shed snow adequately, and they can collapse in heavy snow-load areas unless the snow is removed.

Hoop House Conditions

Proper ventilation is probably the most important consideration when selecting your prefabricated hoop house or constructing your own. If possible select kits that have ceiling vents that can be opened manually or by solar heat-sensor devises. Heat naturally rises in a hoop house, and with opened ceiling vents and an open door or side vents the heat will dissipate naturally through the ceiling vents without any fans. Solar fans work best in hoop houses, but the wax cylinders must be insulated during the winter or removed or they will crack in the low night temperatures. If you are going to install any type of fans, it's best to install intake fans on the lower end wall of the windward side of the house and place exhaust fans in the upper end wall of the leeward side of the house, opposite your entry door. If you have a small hoop house, just install doors that you can open on both ends. Split doors and screen doors are great for the summer time because you can leave the doors open at night. Screen doors keep the critters out and allow great versatility in managing the house temperatures. If you are on the east slope of the Sierra, summer temperatures can reach over 100 degrees, and it's very important to cover your hoop or greenhouses with at least 50 percent shade cloth or add a swamp cooler to your houses.

Ventilation is an essential consideration in hoop house construction.

Here at Sierra Valley Farms we keep the 6 mil plastic on year-round, and just add a 30–50 percent shade cloth over the film from late June to mid-September to cool the houses for nursery and vegetable production. On the other extreme, your tunnels must be able to be airtight at night and early mornings to save those tomato and pepper plants in late September, so you must be able to seal your end walls and secure your films tightly to the baseboards.

Greenhouses
So what's the difference between a greenhouse and a hoop house? Greenhouses are usually classified as being permanent structures, with temperature controls (fans, thermostats, heaters, humidity control, and/or automated irrigation), and are fabricated in an A-frame design, constructed with rigid, UV-rated polycarbonate panels. For commercial production, they are state of the art and expensive. For the higher elevations in the Sierra, if you plan on propagating your own starts and growing vegetable transplants, annuals, or

perennials year-round, a greenhouse is a must. There are prefabricated models of all sizes. A greenhouse should be built separately from your hoop houses; it is a comfortable place that can contain a propagation table (heating cables), potting table, and raised tables and benches to place your flats and pots on. It should have a gravel floor and a nice concrete walkway for ease of cleaning. You must have electricity in your greenhouse for lighting, heaters, and for your irrigation misters and timers. If you can add a small tankless water heater under your basin or to a water storage tank, you can heat your cool irrigation water; it will help to germinate your seedlings during the cold times of late winter and spring.

Geodesic Domes

If you want to grow food year-round above 6,000 feet and in a heavy snow-load area, you probably want to consider a growing dome, or geodesic-style greenhouse. This stucture has been used with success in Truckee and South Lake Tahoe, California. The design of a geodesic dome can withstand the harshest winds and the heaviest snow loads, along with allowing the most efficient use of sunlight within 360 degrees of all sides of the structure. The growing dome uses a variety of features to naturally heat and cool the dome, including a water tank that acts as a heat sink, solar-activated vents, and an innovative way to keep the soil warm in the winter. The "Sierra Dome," developed by Growing Spaces in Colorado, is a 28-foot dome especially designed and engineered for the Sierra heavy snow load. It can handle up to 250 pounds per square inch and withstand 110 mile per hour wind gusts every 3 seconds. (For more information on Growing Spaces visit www.geodesic-greenhouse-kits.com.)

Garden coordinator Michelle McLean is the co-creator of the Sierra House Growing Dome Project, a parent-led educational program that creates ties between local children and their food systems in South Lake Tahoe. Students at Sierra House Elementary School grow spinach, lettuce, herbs, strawberries, flowers, carrots, radishes, chives, and amaranth in their growing dome. Even in the depths of a cold Sierra winter you can grow food for your family and community, Michelle says. Succession planting, maintaining soil temperatures, monitoring insect issues (they use compost tea every two weeks), and growing food vertically as well as horizontally can increase the produce to feed several families or a small community from one 16-foot growing dome alone. The Sierra House students plant every vegetable; they harvest that food to eat in class as a community or to share with others through their salad bar.

Kids growing their own food in the Sierra. How great is that?

Photos courtesy of Jonathan Cook-Fisher.

The Sierra House Growing Dome in South Lake Tahoe teaches local schoolchildren about their food system while growing delicious food.

Sierra weather is full of surprises! Season extension structures can help.

CHAPTER 8
MANAGEMENT OF SEASON EXTENSION STRUCTURES

The key to managing season extension structures is to grow crops that have similar growing requirements and manage the elements of heat, cold, soil, moisture, and humidity so that your crops thrive and produce well. A good rotation of cold- and warm-season crops helps minimize pest problems along with helping to build the quality of the soil structure.

TEMPERATURE

It is very important to monitor the environment within your structures. You should have a soil thermometer to check the warmth of the soil at least every 2 weeks. A high–low thermometer should be placed about 3 feet above the ground to monitor morning lows and afternoon highs. Record all your highs and lows at least once a week; these records are very valuable year after year to develop specific trends in your gardens and on your farm. You don't want to let your houses get too hot in the afternoons and too cold in the mornings.

Here are some standards that must be adhered to as far as the effect of temperatures on plants:
- 115 degrees is the upper limit. Plants then die.
- 95 degrees is the maximum temperature at which plants grow.
- 75 degrees is the optimum temperature for plant growth.
- 55 degrees is the best average day temperature to keep hoop houses and greenhouses for late fall, winter, early spring growth.
- 35 degrees is the minimum air temperature indoors. No freezing.
- 15 degrees is the lowest temperature to harvest winter crops under row covers.

The rule of thumb for cool-season crops is 35–40 degree lows and no higher than 75–80 degree highs inside the structures. For summer crops, it's 50–60 degree lows and 90–95 degree highs. The key to managing these temperatures is ventilation. You must be there to open and close the doors, vents, fans, and/or heaters, and if you're not you must have an automated thermostat system that will do it for you or you will not have a pleasant experience growing crops in season extension structures. When growing warm-season crops in late spring through late summer, you want to manage your structures between 85 and 95 degrees and not let your summer crops get below 50 degrees. It's very important that you make sure that your solar structures have enough ventilation to cool down in the afternoons. In July and August if daily temperatures are above 90, it's best to open all vents and doors by 9:00 a.m., add some 30 percent shade cloth, and then make sure you shut them down after 8:00 p.m. to prevent an unexpected early morning frost.

WATER AND FERTILITY

Water should be available to the houses year-round. A deep year-round water hydrant (a water faucet system that is buried 2 feet below ground as an anti-freeze water faucet) should be placed in the greenhouse, and a washbasin. Water management is a major component in managing hoop houses and greenhouses. The soils dry out faster in the summer, and don't dry out during the winter. Monitoring and observation is a must; physically going out every morning and evening and feeling for soil moisture is essential, along with understanding the structure of your soils (sand, silt, clay, loams). The objective is to have a nice loam soil, high in organic matter, and grow a diversity of crops with multiple plantings.

Fertility is the key to good crop rotations. You must keep adding composts and manures or fertilizers at each planting to maintain the essential nutrients for vegetable production in a hoop house. Growing vegetables, herbs, flowers, and strawberries intensively in hoop houses can deplete the soils of its nutrients in a short period of time. Remember the biomass that you harvest has taken the nutrients from the soil and has to be replaced into the soil for the next crop. So fertility inputs of manures, compost, compost tea, humus, and organic fertilizers must be part of your program.

Some amendments that I add to our hoop house production are alfalfa meal, feather meal, or cottonseed meal, along with compost from our hoop house compost bins, and 2–3-year-old composted horse manure. I strongly recommend that you set up a compost pile or bin in each of your hoop houses and compost all your house clippings, weeds, and so on. The heat of the hoop house will accelerate the decomposition process and in return add heat to your hoop house. After every crop is finished we remove all the plants and all crop debris, put them in our compost bin and then add the meals (alfalfa, cotton seed, feather), a quarter-inch of compost and a quarter-inch of composted horse manure and lightly spade it in (or rototill it in) no deeper than about 3 inches so not to disturb the deeper soil structure where worms are doing their work. Then with a small water roller, we roll it out to firm the soil and rake it level, and either direct-seed our greens, winter root crops, and lettuces, or pull rows for our summer transplants.

ESTABLISHING SEEDBEDS

There are two considerations for establishing your seedbeds in either a hoop house or greenhouse. In general, you want to capture as much soil temperature as possible during the day, so create raised beds that will warm up quicker during the shorter days of the season and can be harvested easily from both sides. For spring plantings of cool season crops, you want to prepare your soil at ground level so that your soils stay cooler for growing winter crops longer into the late warming spring season, unless you want to get an earlier start on summer crops—then just use raised beds year-round. I like to run my beds the length of the hoop house and create a walkway every 4–5 feet, so that I can harvest from either side of the beds.

There are many different small seeders available in catalogs like Johnny's Seeds. Most recommend a seeder that will seed wintergreens and root crops at about 3-inch spacing across the width of the 4-foot bed. Each seeder has its own specifications for seed depth and spacing, so follow their recommendations. This intensive system is clearly explained in Eliot Coleman's *The Winter Harvest Handbook*; it is very demanding on the nutrient system of your soils and requires a lot of nutrients inputs between crops (manures, organic fertilizers, and composts). I don't like to stress my soils so I seed more sparingly (wider widths between rows), so that my soils can rest between plantings. We direct-seed all of our greens, lettuces, and root crops with a Planet Jr. drop seeder, seeding two rows 6 inches apart, and laying a film of T-Tape irrigation tubing between them. Spacing a foot between the next planting gives me room for harvesting and weeding.

Greens are seeded in rows of two, 6 inches apart, with T-Tape between them.

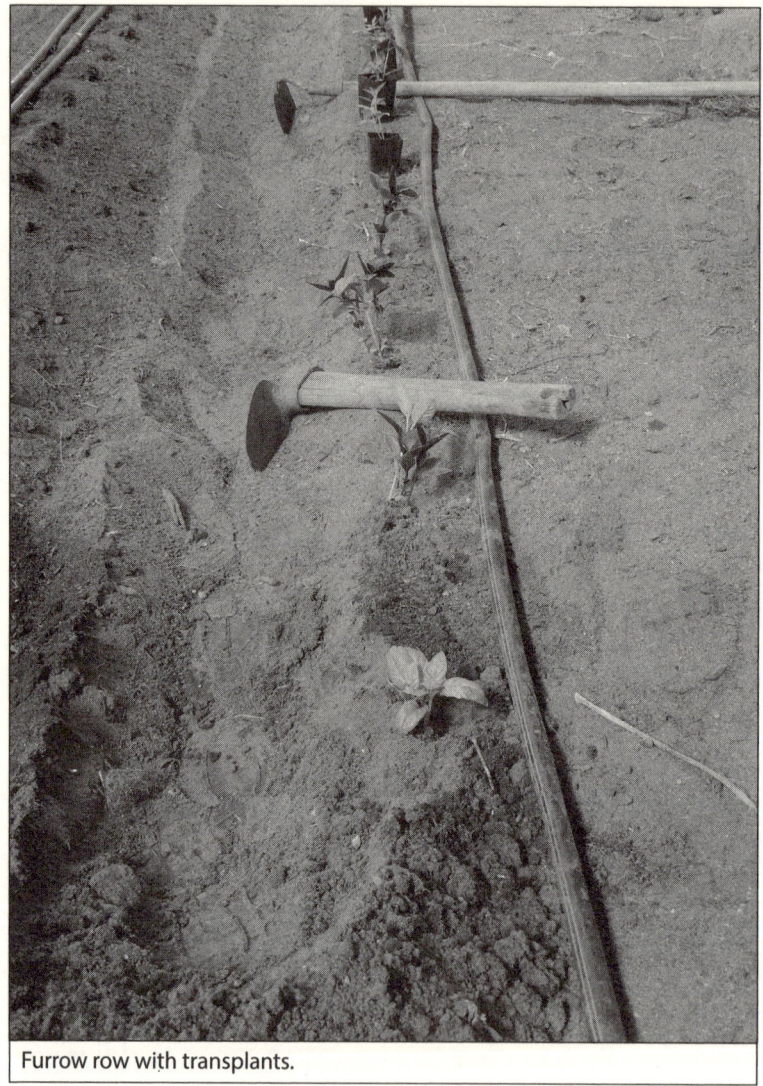
Furrow row with transplants.

For transplants of broccoli, cabbage, cauliflower, tomatoes, peppers, eggplant, and herbs we first make a furrow with a hoe the length of the bed, about 3–4 inches deep, then transplant the vegetable start on each side of the top of the furrow, offsetting them so that they are not directly opposite of each other, about 12–18 inches apart (tomatoes are 24–36 inches apart depending the variety). Then we water with T-Tape for the first 2 to 3 weeks until established, remove the irrigation tape, and flood irrigate the furrows once or twice a week depending on moisture needed. A side dressing or compost teas are then applied into the furrow at flowering bud time to add extra nutrients for fruit production and watered in. Deep-soaking your side dressing and compost teas gives direct nutrients to the roots of your vegetables and prevents any contamination to your above-ground leaves and fruits. Once the furrow is dry you can cultivate the furrow easily to remove any weeds and aerate the soil, then reform the furrow with a hoe and do it all over again. When dry, you can also walk in the furrows to weed when the plants are small, and harvest at maturity.

For propagating cuttings or starting seedlings within your hoop house, you can also consider placing a couple of glass cold frames, dug about 12 inches into the ground, to start your seedlings or root your cuttings. The solar heat during the days will increase the soil temperature and being placed below ground will provide nighttime insulation.

It's always best to keep some sort of vegetation growing in the hoophouses during the winter months. Even in dormancy, plants transpire water and give off heat. Overwintering 1- and 5-gallon plants in pots in hoop houses will absorb heat during the day; barrels of water or stacked water jugs along the sides of your hoop house will also absorb heat.

FROST PROTECTION

As we get into late fall and the winter months it's all about day length based on our longitude and latitude. In California, on Pacific Standard Time, we are pretty lucky here in the Sierra. Based in the northeastern part of the state, we're along the 39th parallel (latitude) and the 120th meridian (longitude). So what does that mean? In layman terms, the summer solstice (June 21) is our longest day (14 hours and 54 minutes), and the winter solstice (December 21) is our shortest day (9 hours and 26 minutes)—everything else is in between. With that in mind, you want to utilize as much day length as possible, to use the sunlight and solar heat that is available. During the cold seasons of the year you want to allow as much light in for the plants to make photosynthesis and capture as much heat as you can to warm up the soil. To enhance that, adding floating row covers for frost protection will insulate the warm soil temperatures, save the heat throughout the night and early mornings to protect plants from freezing, and enhance plant growth.

About row covers: There are many different bonds (woven fabric) that give insulation to plants. The key is to allow enough light through the fabric, yet prevent the plants from freezing. Fabric bonds come in transmissions of light percentages and can be purchased through online greenhouse catalogs like Farm Tek, or Peaceful Valley Farm Supply in Nevada City.

There are usually five types of covers:

    1. 90 percent light: usually for an insect barrier during germination

    2. 80 percent light: the first row cover for light frost, saves about 4 degrees

3. 70 percent medium-grade row cover saves about 6 degrees (good for spring and fall)
4. 50 percent heavy-grade saves 8–10 degrees; most commonly used December–February
5. 30 percent fabric/Typar saves about 10–12 degrees; it is the heaviest grade for over-wintering dormant crops.

No matter what fabric you choose, it's always best to remove the row cover at sunrise to allow as much light and heat in as possible on your crops, and put it back on at sunset. It's definitely a labor of love.

CROP PLANNING

Crop planning is a very important aspect of successful hoop house gardening. I spend a lot of my time during the winter planning what I'm going to grow in each of my five hoop house and greenhouse combinations. I want to have crops to harvest every month of the year, and most times it's the same crops—cool-season crops adapted to my cold climate here in Sierra Valley.

> Staggering your plantings will help you better manage your crops, give you a longer season, and reduce pest infestations.

I have developed a planting schedule of the crops that I grow year round (see Appendix A). I'll begin with January. In summary, I amend all my seedbeds with about 1 inch of compost, 2–3 inches of horse manure, add about 50 pounds of pelleted turkey 4-6-4 formulation (per 10,000 square feet of seedbed), and spade it into the seedbeds about 6 inches. I then lightly rototill them into the beds and hand-rake them level, getting them ready for planting. In February, I begin to plant a rotation of cool-season crops (arugula, kales, Swiss chard, spinach, mizuna, and radishes) in each of my four nonheated hoop houses, staggering the plantings by a week to 10 days in each. These plantings will be ready for harvest starting around March 15 and continue well into May with cut-and-come again crops (after you cut them they regrow to harvest again) like arugula, kales, mizuna, spinach, and chard. After seeding, I cover the crops until about March 1 with #50 Agribon frost protection blankets until they are about 3 weeks old, then begin removing them when the inside temperature lows rise to 28 degrees and above to harden them off. I don't open the hoop house doors until the indoor temperatures reach 100 because I want the soil to capture and store as much of the day's heat as possible and provide a heat source to offset the nightly lows.

When my crops are about finished producing in May I usually use summer as my resting period, and take one hoop house out of summer production. If you have only one hoop house then you'll want to remove half of your hoop house crop and begin preparing your seedbed and planting your summer vegetables; after 2 weeks, remove the second half and plant your summer vegetables. Staggering your plantings will help you better manage your crops, give you a longer season, and reduce pest infestations.

I grow a variety of summer crops, and I especially like keeping a border of perennial flowers and herbs around the inside of my hoop house that I can cut from March through October while I plant and harvest summer annual crops. I try to plant crops indoors that I can't grow outdoors, like basil, sweet and hot peppers, cucumbers, determinate tomatoes, cilantro, and cherry tomatoes, because of the danger of monthly frosts throughout the summer.

As for late summer through winter, I grow cold-season crops and monitor my hoop house temperatures closely, keeping an eye on when the inside temperatures drop below freezing and when the afternoon temperatures rise above 95 degrees. As the outdoor temperatures drop, I want my crops indoors to acclimate to the cooler temperatures because they become hardier over time and can withstand cooler temperatures as winter approaches. Yes, they may grow a little slower, but I won't get a total kill with a hard dip in temperatures.

When the temperature drops in December, row covers are placed over seeded crops until they germinate.

If you decide to plant winter crops, it's best to direct-seed all of these crops into raised beds versus directly into the flat soil, because raised beds with rock or wooden borders will absorb heat and keep your soil temperatures warmer. When the outside temperatures get below 20 degrees, I add my wire hoops to the crops spaced out every 5 feet and apply a #50 Agribon fabric over my winter crops inside the hoop house. If it gets below zero, I add a large blanket of 50#Agribon fabric or a large piece of my 6 mil plastic and cover all my existing individual row covers. I like to seed a second crop of cool-season greens like Siberian kale, spinach, mache, or Swiss chard (the most cold hardy of all the greens). When the temperatures get down below 20 degrees outdoors, in November or even December, I place the row covers right on the ground over the seeded crops until they germinate (7–10 days), then after a week I add the hoops to raise the fabric above the new germinated seedlings. You must be committed to pulling the covers off every morning and putting them back on at night, because the plants need all the light and ventilation they can get. It is very labor intensive. If it gets down to 10 degrees or below zero I run a set of Christmas lights attached to the wire hoops under the Agribon fabric at night to give a few degrees of warmth to prevent a major kill. Young plants are very adaptable to the coldest temperatures that you can imagine, and many times will just sit there until the Earth starts warming up in late January. Seeding that early crop in November or December will give you an early harvest beginning in February instead of late March.

Plants are truly amazing if you let them adjust. During January 2013, I had 24 consecutive days below zero, with the coldest being −23 with an average high of 11 degrees. The greens looked totally wilted and finished. I cut them back and by mid-February, as the days got longer and warmed up, all my crops perked up and exploded, even my 14-day sprouted kale and spinach seedlings kicked into gear. I was harvesting again by March 1.

The same goes for carrots, radishes, and beets. During that period one of my hoop houses where I just grow root crops totally froze up and I had no water—we had permafrost in the hoop house 2 feet down on the sides that froze our water hydrant. To prevent the carrots, beets, and radishes from dehydration, I opened one side vent so that the brush rabbits could go in and they ate all the tops—without green tops plants can't transpire water and dehydrate the carrot below ground—to prevent the roots from drying out, and when the water thawed out in February I watered them and they sprouted up again and were ready to harvest a few weeks later. Plants know how to adapt; just give them a chance! It is also good practice to start a compost pile in your hoop house in late fall to add heat to your hoop house for the winter months.

In general, I find that no matter what you do, you will get only one harvest between Christmas and the end of January. It's just too cold, the days are too short, and after the one picking it takes about 45–50 days to get another—just go skiing and enjoy our winter wonderland!

Companion plants are an essential part of pest control.

CHAPTER 9
PESTS: THE GOOD, THE BAD, AND THE UGLY

It really doesn't matter whether you're gardening inside or out; you are going to be faced with a challenge controlling all kinds of pests. Insects, weeds, rodents, birds, rabbits, raccoons, skunks, squirrels, deer, and bear . . . even your own house cat. Then throw in the ones you can't see, like molds, mildews, and bacterial or viral infections. They are all around you! Your first rule of defense is to plant a wide diversity of flowering plants that border your gardens and plots to create insectaries that attract beneficial insects, pollinators, and birds. In organic farming and gardening one of the major ways to minimize pest problems is to utilize a companion planting program—that is, selecting plants that complement each other and together in a garden they reduce pest problems. (See Appendix B for my companion plants table.)

Marigolds most famously repel soil nematodes and Japanese beetles from strawberries, potatoes, roses, and various bulbs and they also discourage rabbits from nibbling in your garden. Herbs of all kinds work, because their aromas repel insects. Companion plants may shade an adjoining species, and compete for light, water, or nutrients; their root structure may loosen the soil, or enrich the soil with organic substances and nutrients, thus altering the environments of adjoining plants that benefits them in a positive way. There may also be the affects of excretions, odors, insect repelling or attracting substances, biotic compounds, and so on. These may directly influence the growth of other plants, or alter the population of microorganisms that live in the soil, or be otherwise effective in the crowded world of animals crawling and flying in and around the roots, leaves, and flowers of plants.

WILDLIFE PESTS

One challenge while gardening or farming outdoors at any elevation in the Sierra is controlling wildlife pests like rodents and squirrels; rabbits; midsized mammals; birds, deer, and bear. This may seem a daunting task, but if you set up your gardens to be wildlife-friendly they can help your pest management program by bringing in an abundance of pollinators, beneficial insects, birds, amphibians, and many beneficial predator mammals (foxes, weasels, coyotes, or bobcats). Even the manure of predatory mammals can keep rodents, rabbits, and many squirrels away. If you have livestock like chickens, rabbits, or lambs you may not be too excited about bringing in predatory animals, and I understand that concern. For all wildlife pest problems, if you are a certified agricultural producer you can call your local agricultural commissioner's office and request that a trapper come out and trap a troublesome animal on your property, or apply for a predation permit through Department of Fish and Game to shoot problem game species like deer, elk, bear, and mountain lions.

If you're worried about keeping wildlife from your garden, though, 8-foot fencing is the most effective way, but also the most expensive and least attractive. Other measures include using scare tactics like scarecrows or mechanical moving objects to deter wildlife from your gardens. Sometimes by scaring animals away, you can change their behavior or travel patterns and they won't return. Remove the device after you're done, or move it to new locations quite frequently so they don't get accustomed to it. Other techiniques include flashing tape on your fences or between fence posts, devices like automated motion sensor lights, sprinklers, or statues of predators like owls, eagles, or hawks.

Here are a few suggestions for the repelling and controlling some specific wildlife problems.

Rodents

Managing rodents (mice, voles, chipmunks, ground squirrels, tree squirrels, rabbits, gophers, and moles) is a full-time job in the Sierra, especially if you are in the forest. I don't use any poison baits, not only because I'm an organic farmer, but because of my pets, my neighbors, and the danger of hawks, eagles, and other wild animals being killed by eating the poisoned rodents. I strongly encourage you to welcome animal predators such as nonpoisonous snakes, foxes, owls, falcons, and hawks into your gardens to help keep rodent populations in check. Bobcats and coyotes can eat large number of mice and rabbits, but may be confrontational with your cats and dogs. You'll have to make a judgment call on this one.

One repellent that seems to work on squirrels, rabbits, and chipmunks is PlantSkydd, an organic product from Sweden (you can find it at your local garden centers) that utilizes various dried blood ingredients from bovine species that causes a predator fear among these rodents. It seems to work on the smaller ones, but I've had limited success with deer. Blood-meal may work just as well spread around your containers and gardens. Other repellents that work include red or black pepper sprinkled among your beds. Capsaicin is the chemical compound in hot peppers that makes them hot and spicy. Rodents, squirrels, rabbits, and even cats will get this on their feet and skin and lick themselves, and it won't take them long to figure out where it came from. The discomfort hot pepper causes to the skin, eyes, and nose will do the trick, but you also must be careful to handle it with care so that you don't get any in your eyes. Crushed chili peppers of any kind seem to work best. They are

Pests: The Good, the Bad, and the Ugly

cheap and once dried out they will store very well, and you can have them for the following season.

Ultimately, a barrier is best wherever possible for squirrels and chipmunks. Place hardware cloth 12–16-inches deep around the perimeter of your raised beds or around your hoop house or greenhouse when constructing if possible; you could place a 5-foot fence around the area to prevent them from clmbing over. Even so, somehow they always get in. Rabbits are the easiest to control; just keep your hoop house doors closed and screen your vents or place a simple 2-foot chicken wire fence around your planter beds. Voles and meadow mice are the biggest problem during the winter, because they will travel under the snow and they love to hang out under your nice warm floating row covers and devour everything in their path. A friend told me to use bubble gum, that they can't digest it and they will die . . . well guess what, I think they had a party under the row covers blowing bubbles (wouldn't that be a great cartoon?), because they kept eating the bubble gum, and there even were more of them. What did work was to drill a hole in the middle of a mouse trap, place a 16-penny nail through the hole and secure the trap to the ground and bait it with bubble gum. Place two traps back-to-back and place a tin dome or shoebox over the top with open ends for voles to enter; they can't resist and you catch them every time. Mint is a great deterrent for voles as well; they do not like mint, but be careful because mint is very invasive in Sierra gardens and can take over your beds. Other plant-based repellents that have had some success are lemon or citronella oils, cinnamon oil, and rosemary.

A simple vole trap.

A Macabee trap for gophers.

What's the difference between a gopher and a mole? A pocket gopher is the one that leaves big piles of soil above ground, like in the movie *Caddyshack*. They dig tunnels 12–18 inches below ground, storing the soil in their cheek pouches (that's why they call them pocket gophers) and deposit the soil in mounds above ground. Gophers have two large front teeth and are herbivores, eating the roots of plants. Moles are insectivores and eat insect and grubs just below the soil surface; they are the ones that make the ground surface tunnels that you see going across your lawn, pushing up the soil as they go along looking for insects. They are pretty much blind, dark black with huge wide paws so that they can breast-stroke their way through the soil. Moles don't eat plant roots, but the upheaving of the soil disturbs and kills small seedlings. Flooding or mechanical trapping best controls moles. For the pocket gopher, the Macabee trap work the best; it's a spring-loaded barbed trap that harpoons the gopher in its burrow. Set it in the mornings after the gophers have pushed up a new mound of soil. You first dig out the mound about a foot deep and find the burrow, then attach a small chain or wire to the trap and attach it to a stake. Set a separate trap in each direction of the burrow, add a big clump of green grass or weeds behind the trap and cover it with soil. You will catch the gopher by the next morning. In addition to trapping or seclusion of rodents, a good supply of farm cats, owl boxes, and Jack Russell terriers are a good recipe for rodent control.

Skunks

Skunks don't normally cause a problem . . . except when a family decides to move under your house. One winter after a heavy snowstorm they dug under my house while I was away and ran my wife and son out with the most incredible stench they ever smelled. I had to smoke out the house with the wood stove to get rid of it. My wife was not a happy camper! If you do have skunks, they are very docile but nasty on dogs and chicken eggs. Keep all crawl spaces and hen houses sealed from skunks. In the gardens they are attracted to bird feeder litter of seed at the base. It's best to keep the seed off the ground. Application of red or black pepper will also deter skunks from a problem area. It is also believed that skunks don't like rap music, or talk radio, so load up a CD player or Pandora and give them a good dose of Eminem, or Rush Limbaugh (I'd leave if I had to listen to Rush all day!). Worth a try!

Raccoons

Raccoons are very smart and cagey; they are little Houdinis notorious for getting into winter rentals, chicken coops, and, of course, garbage. Keeping your doors secure and garbage away from the house or garden is a must. An application of red or black pepper will help as well.

Birds

Birds (black birds, blue jays, crows, magpies, song birds, starlings, and Eurasian doves) can be a major problem during seedling development in your gardens and during the harvest of berries, cherries, and other fleshy fruits and seed crops like sunflowers.

Decoy plants such as sunflowers can be planted adjacent to your berry crops. Some song birds prefer sunflower seeds to berries, and that will keep them busy while you harvest your berries. Black birds and similar large birds tend to like the big-seed varieties of sunflowers, while the smaller song birds love the smaller ornamental varieties as well as the amaranths. One year we were growing Indian corn at Dad's flower farm in San Jose when I was a kid and the black birds were just devouring the corn, so Dad went out and bought a Zon Gun machine, which is a propane-fired cannon that makes a loud *boom* every 2–3 minutes, to scare the birds away. It worked for the first week, but then the birds became accustomed to it, kind of like living next to the railroad tracks.

Another approach for dealing with birds is to give them an unintentional feeding station. A bird feeder or series of them away from your crops during harvest will give you enough time to harvest your crop and keep them occupied.

Overall, bird netting seems to work the best on tree fruits and berries, but some birds will get caught up in the netting; secure the netting as tightly as possible to minimize this problem.

Deer

Deer are a major problem in the Sierra, especially around Tahoe and Truckee and other populated areas. People love to feed their deer. Some treat them like pets. They feast on your ornamental roses and gardens like candy. Remember deer are browsers; they love the lush growth of spring shoots and flowers on ornamental varieties of annuals, perennials, shrubs, and trees. Ornamental varieties from nurseries are bred to grow fast and have big beautiful flowers and leaves . . . just what the deer love! That's why they eat them to the ground! Whenever possible it's best to plant native plants in your gardens. Native plants have adapted to the "browsers" (deer); they grow slow, and they have small inconspicuous shoots that are just browsed by deer, like simple pruning.

If deer pose a problem, seclusion is the best method—try installing at least a 6-foot fence. The fence can either be wire or plastic netting, which also works very well. Determined deer that may challenge your netting can be baited with a single strand of electrical fencing. Place peanut butter on an aluminum strip and hang it from the electric wire at nose height; this will give them a quick zap and teach them not to challenge the netting.

Other deterrents for deer are aromatic plants like lavender, thyme, and chives that do well in the Sierra planted among your vegetables. These, along with garlic, may have some promise. One program that I use here on the farm is give deer a "victory garden." Plant a small crop of miscellaneous vegetables from your old seed packets away from your gardens and allow them to eat it; they really don't care what it is. It will keep them occupied while you enjoy your garden.

Get creative . . . use flashy colors, wind wheels, and motion scarecrows, something that moves that's out of the norm; a quick nonrepetitive movement will usually spook deer and cause them to change their patterns. You will have to move the devices around every week to confuse the deer. Of course, a good watch dog will work as well, but your neighbors may not appreciate it.

Bear

Those of you in bear-country have your work cut out for you in trying to control bear problems. Seclusion with heavy wire (12-gauge cyclone) construction and electrical fencing is your best bet if you want to be serious about it. Bears are good climbers, so the fence should be 8 feet tall with a top electrical wire (5,000 volts) to keep them from climbing over. Electrical pads or mats work very well; many of my beekeeper friends use these for their hives. The only bare skin on a bear is on the nose and the pads of the feet. Giving them a good zap there will do the trick most times. A farmer's trick is to make a "nail entry mat" . . . take a 3 foot x 3 foot piece of plywood and drive about 20 – 30 16 penny nails through it with the nails facing up and place it in front of your doors or exposed windows . . . their tender paws will not like them. Other tricks that farmer friends have used include hanging very scented soaps from the branches of an apple tree to deter them. If you have a compost pile don't add meat scraps; just add vegetation. Don't leave garbage around; put out your household garbage the morning of your garbage pickup. Be careful not to put out bird feeders. Bears love to play piñata with bird feeders because they are attracted to the seed and the sugar waters. Bears will attack beehives for the honey as well.

Bears are discouraged by the smell of human urine. I have a couple of organic friends that go around peeing on their fence line . . . they swear it works! Well, get a keg of beer and have a party and everyone take a section of fence. If your neighbors might be offended, invite them to the party! Sprinkling a heavy dose of red pepper around the drip line and trunks of your trees and keeping a good aggressive dog will keep them away as well.

For those of you in remote areas, there is a designated bear season in California during which bears can be shot, and a problem bear may have to be taken this way. It is an option; check with your local Fish and Game Department.

INSECT PESTS

The first step is identifying your insects. Get to know which are beneficial and which are the bad guys. You can go online to UC Davis Insect Pest Management (IPM) to identify them. (I reference Ralph B. Swain, *The Insect Guide*, 1948. It's an old relic but easy to follow.) Prevention is the next step; setting up habitats and insectaries for beneficial insects to live and thrive is the best defense against pest insects. For your outdoor gardens, an insectary is a strip planted along your gardens of flowering annuals or forbs (alyssum, bachelor buttons, marigolds, dill, calendula, cilantro, and so on) that can significantly enhance populations of pollen- and nectar-feeding beneficial insects. They are usually planted along with your vegetable crops and turned under at the end of the season. See Table 9.1.

Table 9.1 Common Insect Pests	
Pest	Control
Aphids: Small green insects that suck out the juices of new buds, stems, leaves of a plant. Aphids are usually your biggest nemesis. They begin to appear early in the spring as new growth appears. Aphids follow the sugar cycle in plants as you add nitrogen to get plants to grow; the sugars begin to flow to the lush, tender shoots that are susceptible to aphid damage. This is the time to begin monitoring the movement of aphids. Ants are a good indicator; ants can transport them into your hoop house or greenhouses; once you see ants, it's time to begin treatments.	Sticky strips, insecticidal soaps, ladybugs, lacewings
	con't

Table 9.1 Common Insect Pests (continued)	
Pest	Control
Blister beetles: There are many different kinds in the Sierra and Great Basin. I have them here in Sierra Valley, they come from the hay fields. They are slender and gray in color and can appear overnight in masses. They prefer spinach, tomatoes, squash, beets, and chard.	Knock them off into garbage bags and dispose of them, or use sabadilla dust that is safe to use around pets. As its name implies, the beetle can cause a blister if handled, so use gloves and avoid skin contact.
Cabbage loopers: This inchworm is grayish to green in color and is found under the first couple of leaves of your cabbage and all cabbage members. It's more of a nuisance than a problem.	Hand-pick them off. Birds love them, or you can use *Bacillus thuringiensis* (BT), which is available as a spray at your local nursery. Very safe to use.
Colorado potato beetles: Round, three-eighths of an inch black-and-yellow striped beetles that can do damage to potatoes.	Plant garlic and marigolds amongst the potatoes to repel them; they are easily hand-picked off and squashed. A favorite of ladybugs.
Cucumber beetles: The spotted cucumber beetles tend to be seen more on the drier east side of the Sierra and Great Basin. They feed on squash, melons, cucumbers, tomatoes, cabbage, asparagus, and beets.	A nuisance easily repelled with cayenne pepper or insecticidal soaps. Crop rotations help.
Earwigs: A common household nuisance. They eat everything.	Send the chickens through . . . or, the easiest way to control them is to trap them: Crumple and dampen newspaper and push it into a garden pot. Place it upside down in the garden; elevate it slightly with a rock or stick so the earwigs can get underneath at night. You can then remove them in the morning in a plastic bag, or add them to your burn pile.

Table 9.1 Common Insect Pests (continued)	
Pest	Control
Flea beetles: They are my biggest problem! They like anything in the mustard and cabbage family, or Asian vegetables, especially arugula and radishes. A tiny black beetle—"flea-like"—that hops from plant to plant popping tiny holes in the leaves of young seedlings like a shotgun.	Cover seedlings with insect cloth, or grow these crops in hoop houses. Flea beetles don't like to go under covers. They can be sprayed with insecticidal soap or garlic spray.
Leaf hopper: The leaf hopper is one-eighth of an inch long, lime green and very quick. It is a carrier of curly top, a disease that disfigures beans, beets, lettuces, spinach, squash, and tomatoes.	It likes weeds and decaying debris, so keep your gardens as weed free as possible. Insecticidal soaps help keep under control.
Leaf miner: Can be a pest on spinach, Swiss chard, and kale. These tiny green caterpillars, maggot-like, are pests that tunnel between the leaf layers of the plant, causing a wilt of the leaf. Because they are "under the skin" they can't be removed and are hard to control.	Remove infested leaves and trap them with yellow "sticky" paper.
Spider mites: Spider mites (two-spotted mite) are tiny spider-like mites that develop a small inconspicuous web on the undersides of large leaf plants like thimbleberry, cucurbits family members, beans, hops, and peas. They become a problem as the plant matures.	Use miticide sprays from your local nursery, garlic spray and a heavy oil-emulsion of insecticidal soaps, or ladybugs. They breed in leaf debris and old leaves; keep area around plants clean. Can be a problem in controlled environments like hoop houses.
	continued

Table 9.1 Common Insect Pests (continued)	
Pest	Control
Squash bugs/Caraba bugs: A major pest problem! That's why I don't grow melons, squash, zucchini, or any of the cucurbits family members outdoors; they are hard to control. A black shielded bug that sucks the juices out of the leaves and stems of the plant, and gives off a very unpleasant odor when disturbed.	Hand-pick at the first signs of small infestations, and try to manage them in small populations. They hide under the leaves during the hotter parts of the day and on top in the cool mornings, until August, when they seem to disappear when it gets cooler. These critters like to spend a lot of time on the ground, so a good dusting of diatomaceous earth will scarify their exoskeleton and cause them to die. Eggs are reddish brown and can be sprayed with a heavier oil/insecticidal emulsion to smother the eggs.
Thrips: Small rasping insects that feast on the flower buds of plants, especially roses, and will scar the pods of peas and beans and destroy squash blossoms.	Cayenne pepper works well, as do insecticidal soaps.
Tomato hornworm: A gnarly, ugly beast caterpillar, 3–4 inches long that can camouflage itself to blend in with the stems of your tomato or potato plants. Can defoliate a full tomato plant in a day!	They are easily hand-picked and fed to the chickens; birds love them. *Bacillus thurigensis* (BT) does a good control of them as well. If you have braconoid wasps around them they will lay their eggs on the back of the hornworm and use them for a future meal.
White flies: More of a nuisance than a problem if infestations are kept low. They are kind of hard to control. I just don't plant beans in my hoop houses because they tend to attract them, just as tomatoes do.	A lot of yellow sticky traps or a good shop vacuum works well . . . shake the leaves of the plant and get them flying and suck them up! Or just good old insecticidal soap spray.

The best defense is a good offense. Creating diverse habitats of flowering annuals, forbs, herbs, shrubs, and trees will attract beneficial insects like:
- Black stink bugs or calosoma beetles that lifts their butt and leaves a "stinky" odor (eat a lot of caterpillars and other insect pests)
- Hover flies (looks like a small bee or wasp; eat aphids, leaf hoppers, and mealy bugs)
- Lacewings (eat mites, scale, thrips, leaf hoppers, mealy bugs)
- Ladybugs (eat aphids)
- Parasitoid wasps *(Braconid, Chalcid*; eat aphids, scale, mealybugs, larvae of catepillars, and beetles)
- Praying mantis (eats all insects)
- and other insects that keep the insect pests in balance.

Spiders (arachnids), although not classified as insects because they have eight legs, are very beneficial in the garden, even though most of us don't like them around. Other fun deterrents include lizards, snakes, frogs, and birds. Insect netting and row covers work very well for newly seeded or transplants of vegetables for aphids, flea, beetles and cutworms. Insecticidal soaps and sticky traps work well for small flying insects and aphids, mites, and whiteflies. BT (*Bacillus thuringiensis*) is a pathogen that infects the larvae of most butterflies and moths and works well against cabbage looper, tent caterpillars, tomato hornworm, and cabbage worm. It is safe around pets, birds, bees, lizards, and other animals.

> For heavy infestations it's best to just remove the crop or an isolated plant and burn it, or dispose of it far, far away.

Whenever possible, try to avoid using commercial pesticides. They are toxic not only to the applicator and the environment but also to fish and wildlife. For heavy infestations it's best to just remove the crop or an isolated plant and burn it, or dispose of it far, far away.

INDOOR PESTS

As for insects, I prefer growing outdoors to indoor every time! Any time you have a controlled environment—like cold frames, hoop houses, and greenhouses—you will create a great climate to attract insect and weed pests. It's a real challenge to grow organically indoors without spraying some sort of insecticide, and it takes a lot of monitoring and planning to be successful in your integrated pest management program (IPM). It took me 2 to 3 years to get a balance in my hoop house and greenhouses between beneficial insects and insect pests. They key is observation and monitoring; companion plantings; having a wide diversity of crops; crop rotation; quickly removing infested plants; and creating beds of perennial herbs and flowering plants.

At Cal Poly years ago in the horticulture program I was taught to sterilize everything and keep the greenhouses sterile, weed free, and spray, spray, spray. But all I kept seeing were insect pests, because there wasn't an environment for beneficial insects. The key is to learn how to attract the good ones (insects) and discourage the bad ones, and observe what's going on in your hoop house or greenhouse operation. That's why I went organic. I could see that every time I sprayed I was nuking everything, and the only ones coming back were the bad ones, so I wanted to create a harmonious environment in the houses so that they could be in balance. In 2000, I got the farm certified organic, and it has really paid off. I've taken my outdoor philosophy of plant diversity and creating harmonious environments indoors. I have created wet-managed environments under my propagation bed in my heated greenhouses for habitats for frogs; created perennial herb beds and a diversity of crops to release ladybugs and lacewings to live, and have wider, half-inch screens to allow lizards to come in late summer and get the grasshoppers and true bugs. I've created my own oasis within these hoop houses and greenhouses so that they can maintain themselves. Garden spiders of all types have been aggressive on aphids and spider mites, along with small psyllid and paper wasps. In addition, I use a lot of sticky paper to trap fungus gnats and white flies and identify whatever else is moving in.

The management of crops is essential for keeping balance in a hoop house or greenhouse operation. I rotate crops in and out every 1–2 months, and if a plant seems to be infested with insect pests or has a disease, it's outta there!

Simple weekly applications of insecticide soap will usually discourage an infestation, and bringing in an annual crop of spring ladybugs works great. For specific controls of individual pests, see the previous section.

WEEDS

Oh, those ever-present weeds! Here on the farm, our philosophy is the sooner the better. In our hoop houses we use solarization. Twice a year, once in spring and once in late fall, we remove all crops down to bare ground, irrigate it and lay a sheet of 6 mil clear plastic over it, seal all the edges, and leave it for 3–4 weeks. This germinates all the weed seeds, and then fries them. It's a great way to remove 90–95 percent of your weeds. Then we top-dress with our compost and amendments and direct-seed or transplant into the beds. Outdoors in our open fields it's a race against time. We want the seeded vegetables up faster than the weeds. The sooner they germinate ahead of the weeds, the better our chance of winning the race. We try and weed out 90 percent of the weeds in the first 2 weeks after germination using hula hoes. We like to plant very close together to reduce competition; mulching with straw or newspaper has limited results, other than holding moisture, and it tends to bring in earwigs, pill bugs, and voles . . . more problems. Sanitation of a clean, weed-free bed works a lot better for us.

Solarization is a great way to remove 90-95 percent of weeds. We do it twice a year on the farm.

Some other suggestions are flaming the weeds (burning them with a torch) when weeds are small; this works very well because you can direct-seed or transplant afterwards without disturbing the soil. Anytime you disturb the soil you bring up more weed seeds. At all costs, do not use Roundup or other herbicides that can persist in the soil. If you want to use a spray, you can make up a mixture of Dove kitchen soap, white vinegar, and sea salt and spray weeds on a hot day—it will kill them back at least for a little while. The key is to manage weeds so that they don't take nutrients away from your crops and don't hinder your harvest. You'll likely never get rid of them, and if you're pulling a 3-foot weed, then you've lost the battle. Removing weeds when they are small is essential.

MOLDS, FUNGUS, AND DISEASES

In the Sierra we are pretty lucky, because we naturally live in a dry climate with low humidity. Usually allowing the dry air to circulate in the hoop house or greenhouse is enough to prevent molds, fungus, and diseases (MFDs). The winter months are our biggest problem, because the humidity can be 100 percent; if you're keeping the indoor night temperatures around 40–50 degrees, you will not have too many problems. MFDs don't like the cold; they prefer moist warm conditions. The key is not to overwater your vegetables during the winter. Check soil moisture weekly; in most cases, you only need to water once a week, sometimes only every 2 weeks because with high humidity and cool temperatures your soils will not dry out. There is very little transpiration going on, and the hoop houses create dew that replenishes the soil moisture.

If you use well water, some molds and mildews can occur on the soil and on the leaves; a mixture of 30 percent milk and water will reduce mildew on vegetables, and applications of 10 percent chlorine and water, as well as a 50/50 mix of white vinegar and water, will remove molds on hoop house or greenhouse surfaces. It is important to have sufficient aeration throughout your houses at all times of the year and to allow your soils to dry out and then water them. Because continuous moisture and heavy soils are not compatible, they create anaerobic conditions that allow molds and fungi to grow. Continue to work composts, aged manures, and organic matter into your soil to allow for ample aeration and the addition of healthy microbes.

In general, as temperatures get warmer in spring and summer there is very little danger of fungus and molds. As you enter fall and winter, the humidity rises and temperatures cool and you must pay more attention to your watering schedule. Mildews begin to show and some soil molds may begin to appear. Just don't overwater your crops.

CHAPTER 10
A FIRESIDE CHAT AND GOOD NIGHT

Well, I hope you enjoyed my book, and that you will return to it. Again, my intention is to encourage people to garden in the Sierra, and, for those really interested, to become farmers who can create diverse, viable, sustainable farms and produce food for themselves and for our local communities. That is what this book is all about.

Although I've concentrated on how to grow food such as fruits, vegetables, herbs, and nuts, I have not talked about the aspect of raising animals for food, because that is not my expertise. Animals are a very important element to complement your gardens or farms. Their grazing and feeding habits help reduce biomass for fire danger, reduce insect populations, and their manure is important for your compost pile and fresh fall applications around your berries, trees, and vegetable fields.

> You must look at your garden or farm as a living organism that continues to live, consume, and die, participating in Mother Earth's life cycle.

Wherever and whenever possible, incorporate as many different types of crops into your garden (cool-season and warm-season vegetables, fruit crops, herbs) and consider a livestock element that will complement your gardens (recycling and adding nutrients) such as poultry, sheep, goats, horses, cattle, or hogs.

You must look at your garden or farm as a living organism that continues to live, consume, and die, participating in Mother Earth's life cycle. You must continually nourish your farm and gardens with composts, cover crops, and animal manures, and add to your plant diversity with annual and perennial crops, along with flowering shrubs and trees. Creating habitats for wild and domestic wildlife is essential to keep the balance of nature in sustaining our gardens and farms for future generations.

We enjoy Mother Sierra's natural beauty as we hike to her mountain lakes and swim, cross-country ski, snowshoe, fish, rock climb, mountain bike, and do other outdoor activities. Gardening or starting a small farm is a fascinating way to bring wildlife to you instead of going out looking for it. Once you establish your gardens and incorporate native plants, insectaries, buffer strips, and your hoop house and gardens you'll be amazed at the insect life, bird life, and wildlife they attract. Yes, they will challenge you for that produce, but it's a "win-win" for everyone. There is nothing like the joy of young kids playing in the dirt and picking that first carrot, zucchini, or radish, or tearing down the straw hill to get to a treasure hunt for potatoes in the fall. You'll be amazed what a small garden can provide for three or four families. You'll be giving away tons of squash and tomatoes to your neighbors.

What you really will enjoy are the "flavors of your labor"—there is nothing like the taste of fresh-picked spinach, sweet fall carrots, rich creamy potatoes, or that juicy just-picked berry, apple, or Heirloom tomato. That's what makes growing a garden in the Sierra very special. And yes, it can be done and done for most of the year.

I leave you with this thought: Whatever you decide to grow in the Sierra, it will be a challenge, but go into mountain gardening with an open mind, and no expectations. There will be failures and successes, and it's a continuous learning curve, but you will always be rewarded with the reaps of your harvest, that much I can guarantee . . . and to all a good night!

Three generations at Sierra Valley Farms: Joey, Gary, and Lou.

APPENDICES

APPENDIX A
SIERRA VALLEY FARMS GROWING SCHEDULES

Succession Planting Outdoors • No Row Covers

	April	May	June	July	Aug	Sept	Oct	Notes
Arugula	DS	DS	DS	DS	DS	DS	DS	Cover with insect cloth for flea beetles
Beets			DS	DS	DS			Thin once established
Broccoli/Raab/Rapini		TP	TP					Overhead water to keep cool and prevent aphids
Cabbage/Head/Napa		TP	TP					Overhead water to keep cool and prevent aphids
Cauliflower/Romanesco		TP	TP					Overhead water to keep cool and prevent aphids
Carrots/Chanteney	DS	DS	DS	DS	DS			Thin for larger sizes, not for baby
Chard/Swiss	DS	DS	DS	TP	DS	DS	TP	
Cilantro			DS/TP	DS/TP	DS/TP			Bolts quickly; will overseed itself
Onions/Green/Scallion/Leeks	DS	TP sets	TP sets		DS	DS		Fall seeding will overwinter for spring

SIERRA VALLEY FARMS GROWING SCHEDULES

Succession Planting Outdoors • No Row Covers (continued)

	April	May	June	July	Aug.	Sept.	Oct.	Notes
Horseradish		TP	TP					Perennial
Kale/R. Russian, Lacinito-All	DS	DS	DS	DS	DS	DS	DS	Young sprouts will get frost damage
Lettuces, Leaf/ Spring Mixes	DS	DS	DS	DS	DS	DS	DS	All varieties
Mizuna/ Mustard/Asian	DS	DS	DS		DS	DS		Cover with insect cloth for flea beetles
Quinoa/Greens			DS	DS	DS			Sterile seed over 90 degrees summer
Radish/ RedWatermelon	DS	DS	DS	DS	DS	DS	DS	Do not thin
Spinach/Tyee/ Bloomsdale	DS	DS	DS		DS	DS	DS	Hardiest of all greens; skip July, too hot
Sunflowers		DS	DS					Ornamental and seed varieties/windbreak
Sorrel	DS/TP	DS/TP	DS/TP					Perennial, will overwinter

Legend: DS: direct-seed; TP: transplant

APPENDIX A
SIERRA VALLEY FARMS GROWING SCHEDULES

Succession Planting Outdoors • Other

	April	May	June	July	Aug	Sept	Oct	Notes
Asparagus	TP	TP						Cover for frost protection
Herbs		TP	TP					Companion and border/host pollinators
Perennial Flowers/ Wildflowers	DS	DS/TP	DS/TP			DS	DS	Border plantings/pollinators/insectary
Native Plants	TP	TP				DS	DS	Border, windbreaks, pollinators
Strawberries/ Everbearing	TP	TP				TP		Border, host for beneficial insects
Dill/Fennel		TP	TP	TP	TP	TP		Fennel is perennial/host for good bugs
Potatoes/Yukon/ Fingerling		TP	TP					
Garlic, hard-neck (HN) and soft-neck (SN)	DS (SN)	DS (SN)				DS (SN)	DS (HN)	Overwinters well

SIERRA VALLEY FARMS GROWING SCHEDULES

	Succession Planting Outdoors • Other (continued)				
Squash/Zucchini/ Melon/Pumpkins		TP	TP		Cover & select ones w/earliest maturity
Tomatoes/ Determinate			TP		Cover & select ones w/earliest maturity
Peas (Snap, Sugar)	DS		TP/DS		Use trellis and away from windy areas
Kohlrabi		TP	TP		Overhead water to cool

DS: direct-seed; TP: transplant

APPENDIX A
SIERRA VALLEY FARMS GROWING SCHEDULES, CON'T

Hoophouse Planting Schedule

	Jan	Feb	Mar	Apr	May	June	July	Aug	Sept	Oct	Nov	Dec	Notes
Arugula		DS	DS	DS	DS	DS		DS	DS	DS			No seeding July and late winter
Basil						TP	TP	TP					Transplant after frost, cover in fall
Broccoli (Raab, Rapini)			TP	TP					TP				Spring and fall crops. Rotate years.
Cabbage (Head, Napa)			TP	TP					TP				Offset crops with broccoli
Cucumbers/Pickling				DS	TP	TP							Grow on trellis system, intercrop w/ herbs
Herbs		TP	TP	TP	TP			TP	TP				Use for borders/host for beneficials
Kale		DS	DS	DS	DS			DS	DS				Spring and fall crops, covered Nov-Feb.

152

SIERRA VALLEY FARMS GROWING SCHEDULES

Crop											Notes
Lettuce (Butter, Head)		DS/TP	DS/TP	TP				DS	TP		Seed and transplant for spring/winter
Mizuna/Mustard/Asian Leaf	DS	DS	DS					DS	DS		Mixture of varieties/collards
Maché	DS					DS		DS			Very hardy variety of French lettuce
Peppers (Hot, Sweet)				TP	TP	TP					Need summer heat/cover in fall
Radish (red, Easter egg)		DS	DS					DS	DS		Good early spring and early winter
Spinach/Sorrel	DS	DS	DS					DS	DS	DS	Very hardy, seed Nov, harvest Feb
Swiss Chard	DS/TP	DS/TP	DS/TP					DS/TP	DS/TP	DS	Very hardy, as a TP or direct seed
Strawberries	TP	TP				TP	TP	TP			Everbearing, or small yellow Italian variety

TP: transplant; DS: direct-seed

(continued)

APPENDIX A
SIERRA VALLEY FARMS GROWING SCHEDULES, CON'T

Hoophouse Planting Schedule, con't

Miscellaneous

	Jan	Feb	Mar	Apr	May	June	July	Aug	Sept	Oct	Nov	Dec	Notes
Artichokes					TP	TP							Borders, cutback, cover, mulch for winter
Celery		TP	TP					TP	TP				Will grow as a perennial if cut above crown
Potatoes				TP	TP								Use raised beds; rotate every 3 summers
Onions, chives		TP						TP					Good along borders, plant sets
Garlic		TP			TP								Softneck great for skypes

SIERRA VALLEY FARMS GROWING SCHEDULES

Tomato/ determinate								TP	TP		Seed and transplant for spring/winter

TP: transplant; DS: direct-seed

APPENDIX A
SIERRA VALLEY FARMS GROWING SCHEDULES, CON'T

Heated Greenhouse Planting Schedule
Night temps 45-50 degrees • Day temps 90-96 degrees
Bottom heat: Germination 36 hours @ 80 degrees

Annuals

	Jan	Feb	Mar	Apr	May	Jun	July	Aug	Sept	Oct	Nov	Dec	Notes
Broccoli/Raab		DS/F	TH	DS/F	H, TO			TP	TP				
Cabbage		DS/F	TH	DS/F	H, TO			TP	TP				
Cauliflower/ Kohlrabi		DS/F	TH	DS/F	H, TO			TP	TP				
Microgreens		DS/ F/H	DS/ F/H	DS/ F/H	DS/ F/H	DS/ F/H	DS/ F/H	DS/ F/H	DS/ F/H	DS/ F/H	DS/ F/H		indoor only
Peppers				DS/F	P/H	TH							
Tomatoes		DS/F	F/P	F	F/H	TH							

SIERRA VALLEY FARMS GROWING SCHEDULES

						Others						
Propagate Native Plants	DS	DS/F	DS/F	H	H	TO	TO	TO	TO	DS	DS	
Herbs	H	DS/F	DS/F	P/F	H	H/TH	TO	TO		DS	F/P	H	border plantings
Asparagus		DS/F	DS/F	P/F	H	H/TH	TO	TO					
Onion/Scallion Sets	H	DS/F	DS/F	H	TO	TO		DS/F	F	H	H	spring and fall sets

DS: direct-seed; F: fertilized; H: hardened off; P: potted up a size or two; TH: transplanted to hoophouse; TO: transplanted outside

APPENDIX B
SIERRA VALLEY FARMS COMPANION PLANTING SCHEDULE

	coriander	dandelion	dill	fennel	garlic	corn
anise	yes					
alfalfa		yes				
potato		yes	yes		yes	yes
beets	yes		no			
tomato			yes	no	yes	no
kohlrabi				no	no	
bush beans				no	no	yes
peas					no	yes
horseradish						yes
radish	yes					
carrots	yes		no		yes	
cabbage			yes		yes	yes
pumpkins						yes
rosemary	yes					
pole beans						yes
celery						
NO	fennel					
YES			lettuce		roses	
curcurbits						yes
borage						
eggplant						

APPENDIX B
SIERRA VALLEY FARMS COMPANION PLANTING SCHEDULE

lettuce	onions	carrots	strawberry	potato	YES	NO
yes			no		beans	
yes	yes	yes	yes		kohlrabi	pole bean
		yes		no	asparagus	cabbage
yes	yes	yes			beets	pole beans
	no	yes	yes	yes	cauliflower	
	no	yes		yes	turnip/radish	
yes	yes	yes			pea/bean	hyssop
yes	yes	yes			chive/radish	
yes	yes		no	yes	sage/beets	
				no	sunflowers	
					sage	
	no				sunflowers	beets
yes	yes				leek/cabbage	
		dill		sunflowers		
	roses		spinach	horseradish		
					beans	
			yes		tomato	
					beans	

Lou Romano disking the soil.

APPENDIX C
STARTING YOUR MOUNTAIN FARM

I've written this appendix for farmers just starting out and for people who may want to dabble in farming. We don't have a livestock operation at Sierra Valley Farms. Even though as a child I helped the uncles with a few summer chores feeding their animals here on the farm, I am no way an expert. My focus in this appendix is on organic fruits and vegetables.

SO YOU WANT TO BE A FARMER

Farming is like no other business—it is a labor of love. You have to have a passion to grow things, work outdoors, be your own boss, and provide a good life for your family, and you're out in the country to live a healthy life. That's the intangible part of your business plan. When I created my business plan 20 years ago, some of it penciled out and other parts didn't. I had to throw it away, go from instinct, and find niches along the way. There are so many unforeseen variables that come in to play when you put your business plan into operation that it remains only a boilerplate of the operation. You have to be able to make split-second decisions about your vision and change the direction of the farm because of major obstructions in capital investments, family matters, equipment malfunctions, crop failures, and market losses. It can make it really easy to call it quits.

But back to the intangibles. Farming is a commitment to the cause of growing organic, healthy food for our local communities. It comes from within, not just being able to pencil it out on paper, or knowing you made only made $8 an hour last year. If I had to put wages on my time spent farming it's probably only pennies per hour. When you live on a farm, whether you are starting out or not, the farm never leaves you; you are on call 24/7. When crops are ready to harvest, or need to be planted, weeded, or irrigated, there's no going to Burning Man for a week or the High Sierra Music Festival for 3 days, and there is no paid vacation. It's 7 days a week of making hay while the sun shines. Too many times over the past 20 years I could have called it a career and moved on, but I stuck with it, and got through the tough times somehow. It's a lot of hard work and sacrifices, and many times the reward is not in dollars, but in self-worth, pride, and knowing you are doing the right thing. There is no occupation in life more important than providing healthy and wholesome food for one's neighbors and local restaurants and stores. Successful farmers don't look at farming as work. As the saying goes, *work* is a four-letter word for something you don't like to do. Farmers see it as a way of life, something we

In this photo of 65-acre Sierra Valley Farms, you'll see several of our outbuildings.

have to do to sustain ourselves on the land. There is no price to be put on this lifestyle, but there is a deep satisfaction.

All of us farmers who are still around have stuck with it because it's our life and blood, and we wouldn't have it any other way. If you are in for the long haul or just supplemental income or summer money, that's okay. Just be aware that if it doesn't pencil out on paper, it can be a viable option socially and emotionally, an alternative to a dead-end job and the stressful world we live in. For those of you who really want to read about the nuts and bolts of three generations of successful farmers, our trials and tribulations, pick up a copy of my first book, *Why I Farm: Risking It All for a Life on the Land*. It will give you plenty of insight into the life of a farmer. You can then decide whether this is the life for you.

Planning

To be successful in any business or venture, planning is essential. Farming is no different. There are critical elements of a business plan that are crucial for the start-up phase of a small farm. Remember that business plans in farming are very different from those of other businesses and have different outcomes.

The great thing about a business plan for a farm is that you can create it to grow anything. Start with a vision: include your values, goals, and objectives. Start with a simple mission statement, a sentence or two. "My farm will have the best variety of apples in the county," or "My organic farm will grow nothing but heirloom varieties." Take your vision and build on it, and be true to your mission statement. It will give you guidance and self-worth from the start and inspire you to the end result.

After you create your mission statement, master-plan your property and map out your fields. This is very important, because next you will have to lay out your infrastructure, structures, and hardscapes (for example, walkways, paths, and driveways). If you don't own land and are starting from scratch, there are a few options. If you are looking for land and have the ability to finance, look for an existing farm that's up for sale, or one for lease; that will give you a head start. If you are not sure about the farming thing, whether you will like it or not, then look to lease, rent land, or try a crop share with your neighbors—going in on it together, putting in a crop, sharing equipment and irrigation, and splitting the profits.

Now that you have your vision, direction, mission statement, and goals and objectives you need to put together the pieces of the puzzle. This involves installing the infrastructure of utilities, water lines, buildings, and structures. Try to centralize your utilities (water, power, gas lines, and sewer) in one area so that you can trench them all together or in proximity to each other. Try to design a solar and wind power element into the planning and implementation phase of your utilities. If you are off the grid, it is essential that you design your farm with your power system as the priority. Lay out your solar systems or create your out-buildings in the sunniest areas, away from trees and in a southwest-facing direction. If you have any ridge tops or elevated slopes, consider wind power, because winds tend to blow mostly during the night before and during spring and winter storms. If you have permanent or intermittent streams, consider using small hydroelectric units that are very efficient in providing localized power. Even if you are on the grid, consider all of the above and pursue state and federal rebates through your local power provider.

Water

Water is the key to all your crops, indoor and out. Without adequate and good-quality water there is no use farming. There is probably nothing more important than planning out your water sources and irrigation system for your farm. First, it starts with your water source. Do you have well or domestic water, or pump from a pond or water collection system? Next, a simple water test will reveal the quality of your water. Are you high in minerals or sodium? Do you have hard water (high pH, calcium, and magnesium), or soft water (low pH, calcium, magnesium)? Know what you have to work with. Hard waters high in calcium or sodium can calcify and clog up your drip systems, and soft waters low in pH can inhibit the ability of plants to take up necessary nutrients through their root hairs. Only a water test can tell you that.

Probably the biggest constraint you will face in the Sierra is the limited availability of water. As you get more remote and higher in elevation, you will have to dig deeper for water to get through the granitic layers of bedrock to the water caverns. Most wells in the higher elevations are going to produce lower volumes of water, less than 10 gallons per minute (gpm), and many may only produce 2 to 3 gpm. In that case, you may have to design your systems to include water storage tanks. You can feed your water tanks with solar pumps, windmills, or electric pumps. All your well houses must be well insulated from the winter cold, with electricity so that you can add heat tape, heat lamps, or small area heaters to keep at about 50 degrees. All pipes should be wrapped in insulation. All agricultural water storage tanks should be separate from the domestic water systems and tanks, and located on the higher parts of your property. They should be pump fed or gravity fed to pressurize your irrigation systems, and be able to be shut off, drained, and winterized for the winter. Your low-volume irrigation systems must have pressure regulators that will keep them between 30 and 50 pounds per square inch (psi). If you need water storage for winter operations, it should be plumbed into a heated well house or building to prevent winter freeze damage. Your local well drillers can help you with that, or your Natural Resource Conservation District (NRCS) regional offices.

You must design your water systems according to the crops you are going to grow for each season; first, to estimate your water demands, and second, to lay out your irrigation systems so that they can be isolated for individual crops, or shut off and winterized for those areas not being used in the winter time. In the Sierra there is always danger of frost and winter permafrost, so try to lay out a manifold of main lines throughout your production areas so that you have water access to every part of your farm. You can never have enough water lines. Even if you don't plan on using one section of the farm, plumb it anyway because it's much cheaper to do it all at once. Put all your main lines at least 2 feet deep and install drains and isolation valves so that you can shut off areas that are not going to be in use. All lines must be able to be winterized with manual bleeder valves, or with automatic drains below the frost line, and use deep-freeze hydrants in buildings and greenhouses that will be unheated.

As I sit here in Sierra Valley on this thirteenth day of January, typing away, I look at this morning's temperature of 21 below zero, and say, "Why didn't I put those irrigation mainlines deeper than 18 inches, and plumb them directly into the greenhouse with deep freeze hydrants?" I wouldn't have been frozen up this morning with no water to water my plants.

When designing your irrigation systems for your crops, in most instances low-volume drip irrigation will be the preferred choice. There are a number of different water application selections to choose from, depending on your water source and crops, from emitters to microsprinklers. Remember no matter what your water source—whether it's a well, domestic source, or a pond—all your systems must have a filtration system to keep these fine emitters and orifices free of obstructions, and all systems attached to a house well must have a backflow device to prevent water contamination. See your local environmental health departments for their requirements and permits.

When designing your irrigation systems, I like to use DripWorks, out of Willits, California (www.dripworks.com), or Peaceful Valley Farm & Garden Supply (www.groworganic.com), in Nevada City, California. They can help design the system to meet your specific needs.

Water Management

Once your water infrastructure has been established, your main concentration as a farmer is managing water and efficiently applying the water resources you have available to your crops. One thing that sometimes goes unnoticed in the Sierra is water temperature. When I first started 22 years ago here in Sierra Valley, I built a greenhouse and two hoop houses and began to grow native plant seedlings and vegetable starts. I designed an elaborate hot water hydronic system to heat my soil beds for germinating seeds, but as the misters came on, or as I hand-watered the flats, I noticed it was taking me 2 to 3 weeks to germinate simple vegetable seeds that should only take 5 to 7 days. I finally realized that my 300-foot well water was 38–42 degrees. I was defeating the purpose of heating the soil to 70–78 degrees by adding 38–42 degree water. So, in my greenhouse I added an old steam boiler tank, painted it black, and tapped in a hot water source from my water heater—this became my mixing tank. I then was able to keep room temperature water available for watering my indoor crops. This is a must if you plan on growing during winter or spring.

There are a number of ways to heat water: You can run water into black 55-gallon drums, allowing the water to heat in the greenhouse or hoop house, or use solar water systems, tankless water heaters, or a mixing tank system similar to what I did. Selecting the right sprinklers and applying water to your crops in an efficient and effective manner will be one of the biggest challenges you face as a farmer in the Sierra.

Facilities

The main facilities you will need for a farm include:
- A shop or barn to house equipment and supplies
- Garages or "lean-to sheds" that house equipment, vehicles, and machinery, so that you can work indoors on projects
- A packing shed to clean and package fruits and vegetables
- A food storage facility, either a cold box, root cellar, or walk-in cold storage to house all your fresh-picked fruits and vegetables
- A bulk storage shed to house all your packing boxes, packaging, and bulk dry storage supplies
- An employee building for employee or worker restroom and showers, a break room, meeting room, kitchenette, office and phone, and/or record keeping
- An insulated well house—it's very important that your well house be well insulated so that you have water available during the cold winter months
- Season extension greenhouses, hoop houses, or cold frames
- A propagating greenhouse to germinate your vegetable starts early in the season
- Season extension hoop houses or greenhouses to grow vegetables during the spring and fall seasons

These are the main buildings of concern; remember, you can never have enough buildings. Make sure that you have at least one large building that is well insulated and heated, with a concrete floor, and with power and electricity. Nothing is worse than trying to work on projects or fix things when it's 10 below zero in a tin building. When planning your out-buildings and structures make sure you centralize them and give enough room for snow removal operations, and try to make your walkways close to plowed driveways for ease of snow blowing or shoveling.

As you can see, there is a lot to plan for, and it's an ongoing challenge year after year. You start with what you have to work with and can afford. Remember you don't have to buy everything new. Go to auctions; frequent Craigslist and yard sales. If you are a good mechanic, or know a good one, buy used equipment in good condition and inspect it. If you are looking for tractors, find ones that have 1,000 hours or less in operation. Tractors are based on hours of use instead of miles. Tractors that have more than 2,000 hours are usually pretty beat up; it's like buying a car with 200,000 miles. Just buy the essential equipment that you need first, and then every year add to your arsenal as you can afford it.

Financials and Budgets

Resource analysis is next. What resources do you have at your disposal? Consider savings accounts, CDs, IRAs, credit, personal loans from family and friends, and side jobs to help fund your farm. Pool all your resources together.

You need to begin to layout your annual budgets and plan to finance them. A word of warning, even if you have good credit: Lending institutions like Bank of America and Wells Fargo consider farming ventures high risk and very seldom will loan you money against a farming venture. You may have to get a home equity line to finance it, or try your local banks or credit unions. The amount you need will vary depending on the size and extent of your operation.

There are two financing budgets you will need to create: **capital budget** and **operating budget**. The capital budget includes the capital outlays you will need for fixed items such as land, wells or ponds, structures (houses, out-buildings, greenhouses), infrastructures (irrigation lines, sewer, and utilities), large equipment (tractors, implements, farm vehicles), and hardscapes (driveways, pads, roadways). You could fund your capital budget through refinancing your home or opening a farm loan, personal loan, home equity loan, or creating a farm lease or rental.

Your operating budget will include whatever you need to put in the ground and anything to do with operating your farm business. Remember, if you plan to live on the farm and it is your main income, pretty much everything is deductible except vacations. Your budget should include everything you can think of: seeds and plants, amendments and fertilizers, pest control, tools and supplies, fuels, utility bills, office supplies, marketing and advertising, travel (to markets, community-supported agriculture [CSA]), maintenance of equipment, labor, insurance, and medical bills.

You'll need start-up cash to purchase seeds, plants, labor, materials, and supplies. How much money will you need to start up all your crops for the season? In general, you will need at least 40 percent of your overall operating budget to begin the season.

Labor

Labor is the most expensive part of farming. It can make you or break you in one season. If you don't budget your labor force you will spend your entire operating budget very quickly. What is your time commitment to the farm? Are you going to do it yourself, or hire a manager or foreman? What size crew do you need? Do you need them full time or seasonal, just when the crops are in? Do you plan on paying workers compensation and health benefits, or use a labor contractor? If you have orchards of berries or fruit trees, you will need pruning crews, spraying crews, and harvest crews. Are there people available in your area who have knowledge of all these tasks, or are you going to do it with interns, family, relatives, and friends? Do you have housing on site for farm interns or labor families to stay in? Who is going to do your marketing and sales of goods, attend farmers markets, or put together your CSA baskets and deliver them? Last, what wage or arrangement are you going to pay them? By the basket, bin, hourly, or by the crop? All of these questions must be answered before you try to build an operating labor budget.

Equipment and Supplies

This is your "catch-all" budget. Pretty much all the main supplies you will need on the farm: hardware, wire, irrigation supplies, lumber, metals, fencing and netting, small power equipment and maintenance parts, tires, repair parts, small hand tools, electrical supplies, and construction materials.

Crop Plan/Profit and Loss

You must lay out a plan for every crop. It should contain the field acreage, location, season, crop rotation, cover cropping, how much seed or plants needed, inputs needed (fertilizers, amendments, compost), supplies needed (drip tape, irrigation pipes, sprinklers, row covers, harvest bins, baskets, ties, bags), hours needed to prepare the soil, seed, weed, irrigate, harvest and post-harvest, and marketing plan: Where it will be sold, for example? Farmers markets, farm stand, CSA, or restaurants? After all is said and done, "Did you make a profit?" Assess the value and viability of this crop to see if it's worth growing again, or evaluate what can be done to make this crop more viable or discard it all together from the program.

Financial Contingency Plan

It's very important to have a backup plan. Don't put all your eggs in one basket; diversify your operation so that if something fails you can fall back on something else. What happens if the crop fails or your crews don't show up or your markets don't sell all your products? Equipment could break down, and the well might go dry. Acts of God could decimate your farm in a massive snowstorm, flooding, or a wildfire. How are you going to cope with these losses? What finances are available to make up for the losses?

Miscellaneous Expenses

Other expenses that you will incur include:
- Office expenses (computer, phone, Internet, printing, postage, supplies)
- Legal expenses (fees, charges, certifications, accountants, legal fees)
- Insurance (auto and farm, health, workers compensation, flood, crop, on-farm liability, product liability)
- Marketing and advertising (sales and market expenses, travel, websites, social media, post-harvest packaging, labels, advertising)
- Utility expenses (fuel costs, electrical, gas or propane, water, sewer)

These are the main miscellaneous expenses; there may be specialized ones added depending on your specific situation.

Farm Management

Managing a farm is tough. You have to be knowledgeable in all subjects. Assess your skill and knowledge level of farming. Have you ever farmed before and do you know what you're doing? A farmer must be a jack-of-all-trades—a plumber, welder and fabricator, equipment mechanic, electrician, construction worker, salesman, accountant, manager, and also a crop grower. If you can't perform any of these tasks, whom can you turn to, and who's available to help you? Friends, neighbors, farm advisors, relatives, mentors, and so on, or are you going to hire out? Do you need to attend classes at a junior college, attend workshops or conferences, network with farming associations and other farmers, or even become a working intern on a dream farm that you would like to have? In any case, you need to have a good idea about what you are doing. You don't want to invest all your time and money into a venture that you have no business doing if you are that naïve. Farming is hard even if you know what you are doing.

Personal Time Management

Last but not least, the one thing that is sometimes forgotten is taking care of number one. That is you. Farming is very demanding physically, mentally, and emotionally. It is crucial that you budget time and money for yourself and your family. Take time to have fun away from the farm, even if it's only for a day or a weekend. The farming lifestyle can put a lot of personal stress on a marriage, family life, and personal finances, so it is very important to budget for vacations and personal perks that can relieve the stress of the farming lifestyle. Go out and have some fun.

ALL ABOUT FARMING

What machinery and equipment do you need for your farm? This is literally the nuts and bolts of this appendix. Start by looking at what machinery, pieces of equipment, and tools you will need for each crop task: (1) Land preparation, (2) Seeding and planting, (3) Maintenance and irrigation, (4) Harvest and post-harvest, (5) Sales and marketing, and (6) Farm projects and maintenance.

The Farm

How big is your farm? Is it in acres or in square feet? Do you need a tractor, or small power equipment to work raised beds or acres of fields? Try to select versatile equipment that can be used for multiple tasks by being able to add multiple types of attachments to it. In general, if you have a raised beds and small gardens under an acre you can get by with walking tractors and rototillers, small rollers, planters/seeders, and small power equipment. If you plan on farming 1 to 10 acres you will need a tractor/loader (25–50 horsepower [hp]) with a backhoe option, a plow, disk, rototiller, ring roller, and chisel. These are your basics. Other implements that are labor savors include transplanting machines, row makers, mulch layers, and seeder bars, rolling cultivators, root harvesters, and snowplows or snow blowers.

Crops: Seeding and Planting

Again, it depends on the size of your farm. Acreages versus square feet. Less than an acre (square feet) you should be able to do it with walk-behind seeding equipment, or by hand. For an acre or more you will need a tractor with row markers, mulch layers, bed makers, and tool bar seeders. It also depends on your farming methods and crops, and whether they are fruits or vegetables. With vegetables, some people like to plant directly with seed, and some like to transplant, or both. Each crop needs its own mechanical equipment. For fruit production, you can plant trees by hand, or use a tractor auger or back hoe. How many trees or vines do you have to plant and how big are they?

Crop Maintenance and Irrigation

Once your crops are in the ground, how are you going to weed or fertilize them? Do you have to spray them for pests? Equipment for maintaining crops is mostly weeding hand tools, cultivators, small fertilizer broadcasters, or backpack sprayers. For larger acreages you may consider rolling cultivators, adjustable rototiller attachments, drop spreaders, and boom sprayers. For fruit trees and berries you must consider the pruning element. Do you prune by hand or use mechanized equipment? The same for sprayers; most orchards need to be sprayed with power spraying equipment or backpack sprayers.

Your selection of irrigation methods is an important choice. First, look at the volume of water available in gallons per minute, then decide under what type of irrigation does the crop do best, and last, what type of soil you have. Remember, clays hold a lot of water, so a little goes a long way; sand is the opposite, it can take a lot all at once but doesn't last long. Another factor is that wherever you put water weeds grow, so that's more work for you. A rule of thumb is to put water only where you need it. That's why in most cases drip irrigation is the number one choice for farmers of all sizes. You just have to decide whether you are going to use drip tape, emitters, or microsprinklers. In most cases you will need a minimum of 15 gallons per minute per acre of planted ground under drip irrigation.

Harvest and Post-Harvest

What equipment are you going to need to harvest your crops? Are you going to pick by hand, or use specialized harvesting machines like root harvesters, mechanical fruit harvesters, greens harvesters, and so on? In most cases fruits and vegetables are picked by hand. Equipment needed are usually small tractors, utility carts and trailers, used to transport fruits and vegetables from the field. Washing facilities are set up in most cases; vegetables are washed by hand, and storage facilities must be available in the form of cold storage for perishable vegetables and cool storage for warm-season vegetables, and fruit storage. For fruit production, many are transported in bins to fruit-packing facilities, so those arrangements must be made in advance and the costs budgeted for. These packers also contract with the owners to harvest, pack, and sell the crop and give them a percentage.

Sales and Marketing

The basic equipment for marketing are an office and vehicles. How are you going to sell your product? Is it fresh fruits and vegetables? Are you going to make a value-added product like jams, jellies, or condiments? Are you going to set up a commercial kitchen? Do you need any specialized packaging or bottling equipment, or label makers? Are you going to set up a CSA and need a facility to put together weekly baskets? Do you need delivery trucks, what size, how many? Do you need refrigerated trucks, flatbed trucks, or pickups?

Farm Projects and Equipment

Depending on your skill level and the skill level of your employees, what equipment do you need to fix things, work on projects, and maintain the farm? This list can be lengthy, including power tools such as drill presses, power hammers, table saws, fabricators, radial arm saws, welders, chop saws, air tools, and air compressors; specialty wrenches, mechanics tools, hand tools, and so on. Also consider maintenance equipment to keep up the farm, like lawn and brush mowers, Weed Eaters, snow blowers, leaf blowers, sweepers, and electric and and power pruners.

> Farming in the Sierra is a lot like farming in Europe—Italy, for example. All the farms are small, probably 1–10 acres, terraced along hillsides, with not much flat land.

Because time is money, and labor takes time, the most efficient way to farm is to utilize mechanized equipment to help you with your tasks at hand. You do need to select the right piece of equipment for the right job. The key is to size your equipment to the size of your farm. Just because your neighbor has a D8 massive earthmover doesn't mean it's right to use it to prepare your soil on a quarter-acre lot. You'll do more damage by compacting your soil than if you use a small utility tractor. Every situation will have its own requirements that you must figure out. Of course, unless you're rich, or have a great retirement of endless cash, you must work within your own means, because farm equipment is expensive to buy and maintain. If you are like most of us farmers, it's beg, borrow, or steal (only for a short time . . . we'll give it back), or Craigslist. Like I mentioned earlier, there is a lot of good-quality farm equipment out there; farms go out of business every day, so know what you're looking for and stay active on Craigslist, auctions, hit the yard and ranch sales, and keep in touch with your local tractor dealers so they can let you know when pieces of equipment are being turned in. Be ready to buy when that good deal comes up.

Tractors

I'll begin with tractors. As a kid growing up on a 20-acre flower farm and orchard, and then helping the uncles up here in Sierra Valley with their 3,000-acre hay, dairy, and livestock operations, I was exposed to a lot of tractors. We had them all, small garden tractors, cleat track "crawlers," dozers, loaders, skid loaders, and large-wheel John Deere and Case Hay tractors. There are all makes and models, and it can get confusing deciding which one is best for you.

First look at the layout of your farm and gardens. Farming in the Sierra is a lot like farming in Europe—Italy, for example. All the farms are small, probably 1–10 acres, terraced along hillsides, with not much flat land. The Sierra are pretty much the same, unless you are in a large mountain valley like Sierra Valley, or many areas on the east slope of the Sierra. In my experience, the Romanos have always tried to have enough specialty tractors to go around. In other words, don't have only one tractor to do everything; try and have a few to do many tasks, that way if one breaks down, you have another to take its place, and trust me that will happen. There are four parts of a tractor you should consider when selecting one:

1. *Power size.* How much horsepower does it have to power, pull implements, and lift objects?
2. *Drive train.* Does it have a PTO (power take-off), a spline gear that a drive shaft attaches to power tillers, mowers, and so on, or does it have the ability only to pull implements like disks, rollers, and plows. Is it a hydrostatic drive like a skid-steer tractor that has a "joy stick" that drives all wheels on demand, or does it have 3–4 different gear ranges that you can shift to, allowing 10–12 different speeds so that you can find one that meets your farming needs?
3. *Mode of traction.* What is its mode of traction? There are three basic types: cleat, wheel, or skid-steer tractors.
    - The first, a tracklayer, cleat tractor, or "crawler," as it is sometimes called, has steel or rubber cleat tracks that travel over the ground like an Army tank. Most tracklayers are good only for pulling implements like trailers, sprayers, disks, rollers, plows, or chisels; they have a lot of power, wide wheel base, and a low center of gravity. They are especially good for sloped sites, for orchards that require you to get under the tree branches, and for pulling heavy implements.
    - Today, most farms use the second type: a utility wheel tractor, like a Kubota or John Deere, a very versatile four-wheel-drive wheel tractor that has a PTO, hydraulic power for loader applications, and lifting implements to be transported around the farm.

When it comes to equipment, you need the right piece for the right job.

- The third is the skid-steer tractor, best known is the Bobcat; they have hydrostatic drive to all wheels. These tractors have limited value in field farming applications because their short wheel base and quick turning ability tears up the soil, but they are very efficient in snow removal, soil- and manure-moving operations, and have a lot of attachments like augers, back hoes, tillers, sweepers, snow blowers, rock pickers, bucket attachments, and so on. They are good to have on your farm because they don't tie up your farm tractor doing these other operations.

4. *Versatility*. How many different things can the tractor do for you? Does it have a wide variety of speed ranges, hydraulic capability to add a loader bucket to the front, lifting ability of the three-point hitch to lift different implements on the back, or a mower underneath? Does it have enough power in its PTO to drive snow blowers, or mowers and backhoes? Last but not least is the ease of maintenance. Is it an antique that you can't get parts for anymore, or is it a European- or Chinese-made tractor that you can't find parts for? Stay with farming brands like John Deere, Kubota, Case, New Holland, Massey Ferguson, and Caterpillar.

A 4-wheel drive tractor with a bucket, and a skid steer (in back).

When selecting tractors, look for newer tractors, say less than 1,500 hours of use if possible, test drive them, and have a mechanic look at them. For most farming operations 25–50 horsepower tractors should be able to handle up to 40 acres of farming, and select implements that perform 4–6 (25–30 hp) feet wide, or 7–8 (40–50 hp) foot-wide farming applications, like mowers, disks, rollers, and tillers. The garden tractors have their purpose for mowing rough areas or tilling between rows, snow blowing, and pulling trailers. I use electric golf carts a lot for transporting vegetables from the field, or running back and forth for stuff. All-terrain vehicles (ATVs) are fine for that as well, but they use gas and make a lot of dust, and their attachments in the catalogs are pretty useless in farming operations. Stick with commercial, heavy- grade implements from tractor dealers and auctions; if properly cared for, they will last a lifetime.

Plow

Plows

Single- and double-blade plows are used for initial field preparation or for incorporating cover crops, harvesting root crops like garlic, onions, potatoes, and so on. There are tow-behind models, and three-point hitch attachment models.

Disks

Disk

Disks are probably the most important piece of farming equipment. They are used in the spring to open and aerate soils, and in the fall for incorporating cover crops, manures, and composts. Good for weed and rodent control, and fire suppression in rough areas around the farm. There are tow-behind and three-point hitch models. Select 5–7-foot models for most tractors under 50 hp.

Ring roller

Rollers

There are different types of rollers. Small, flat surface, water rollers for garden tractors roll the soil flat for seedbed preparation in 8–12-foot gardens, while toothed-ring rollers are for breaking up dirt clots for seedbed preparations in field operations, and solid, flat-surface pull-type rollers are used for orchard harvest operations of almonds, walnuts, and fruit trees.

Tillers

There are two types: *front tine* and *rear tine* tillers. Front tine are mostly the garden walk-behind that you can rent; they are only good for incorporating amendments into loose soil in your raised beds or between rows. Most commercial types of tillers are rear tine rototillers. There are walk-behind 5–15 hp rear tine tillers, tractor drawn sliding row tillers, and solid tillers ranging from 4 to 8 feet long. Tillers take a lot of PTO power; try to select 4–6-foot tillers for tractors in the 25 to 50-hp range. Rear tine tillers should be used after a disk, plow, chisel or cultivator, used for seedbed preparation, followed by a roller, has broken up the soil.

Cultivators

There are a number of pull-type cultivators available, from rolling row cultivators for large field operations to individual row cultivators and spring and tine harrows used in pasture aeration and seedbed preparations.

Rolling row cultivator

Chisels or Rippers

These are curved-steel shanks that usually are attached to a tool bar or a steel frame that is 12 to 36 inches in length that are used as "subsoilers," to break up the subsoil beneath the topsoil for aeration and drainage. Cleat tractors work best for these implements, but for wheel tractors keep shanks

Ripper

between 12 and 18 inches in length, with one or two on a bar, spaced 12 inches apart. Applied in spring and fall after disking.

Diggers/Trenchers

A number of backhoes and trenchers are at your disposal. If you don't use them a lot, sometimes it's best to rent small trenchers to put in your water, irrigation, or power lines. Make sure they are capable of digging down at least 24–36 inches. In areas of heavy rock, avoid a trencher and use a backhoe. If you are not proficient with a backhoe, it's best to hire someone, and always remember to call your utility people before you dig to avoid a dangerous situation.

Backhoe

Buckets/Blades

There are a number buckets and blades available for your loader, including straight gravel buckets, 4-in-1 buckets for grabbing limbs and debris, oversized snow and manure buckets, and rock buckets, to name a few, along with different blade attachments like scrapers and offset blades for snow removal.

Mowers

There are four types of mowers: blade, flail, reel, and sickle mowers. *Blade* mowers are the most common. They are like your lawn mower, a series of rotating blades under a protective deck that cuts grass, weeds, and brush. The mowing decks range from 21-inch lawn mowers to 60-inch brush cutters. The mowers can be purchased as tow-behind self-propelled (for garden

Bucket

tractors and ATVs) or three-point hitch, PTO-driven tractor attachment mower deck models. Blade mowers can be dangerous, are known for throwing rocks, and can be damaged if used in rocky areas; these mowers are best for lush green weed mowing in the spring, and for hard packed areas free of rocks and debris.

The next are *flail* mowers. For the Sierra this is the mower I recommend. It is the safest and most versatile. The flail mower has a rotating horizontal shaft with swinging steel blades attached by a chain link to the rotating shaft; as it rotates, it shreds weeds and plant parts in its path. The mowing leaves a rough cut, but because the blades can swing, they can hit rocks, twigs, or pine cones and swing away without damaging the blades or the shaft. Flail mowers do a better job on drier rough areas than on lush green grasses, and what makes it nice is that you can elevate the mower on a three-point hitch and use it to thresh grain and dislodge seeds in your pastures, or to mulch your old vegetables to create green manure to turn under in the fall.

Reel mowers have limited uses. They are basically knife blades surrounding a reel that rotates against a bed knife making a precision cut similar to a scissor. They are used on golf courses mostly, or for hay-cutting operations. Unless you have a hay operation, they serve little purpose on the farm.

Last is the *sickle bar* mower. Some commercial walk-behind mowers use a sickle, or pruning-cut motion, to cut vegetation. Used mostly for large, tall, out-of-control grasses or saplings, or as a sickle bar attachment on tractors for cutting grain and hay crops.

Blowers and Weed Eaters

There's not much to say here. They each have their purpose, and come in all shapes, sizes, string, and blade forms. Basically, blowers blow leaves, and Weed Eaters cut weeds. There are many models to pick from. Wear safety gear when using them, for example, ear/hearing protection, long pants and boots, gloves, and eye protection or safety goggles.

Tool Bar Seeders/Fertilizer Boxes

In larger field operations many farmers use a tool bar, which is a 2-inch square iron bar 6–12-feet long to which they attach a number of precision or direct seeders, along with fertilizer boxes or a spray tank to add water or pesticides. Usually each seeder is assigned to a single row so that you can seed multiple rows at a time. Other implements, such as furrowers, cultivators, shanks, or plows, can be added to the tool bar for other farming operations.

Harvesters

There are crop harvesters like potato and root crop harvesters, grain harvesters or combines, greens harvesters, hay-baling equipment, and fruit and nut harvesters.

Transplanters

There are mechanical transplanting machines that do a great job planting vegetable and tree starts, bulbs, root parts, and tubers directly into the field. They are tractor drawn and multiple transplanters can be attached to a tool bar to transplant multiple rows at a time. One person transplanting and one tractor driver can plant an acre in about 2 hours.

Transplanter

Bed Shapers

Bed shapers are like box scrapers, except they don't have a back blade. They take tilled soil and shape it into a single, or double, 4-foot wide raised bed that is then direct-seeded or transplanted with vegetable starts through a black or red plastic mulch.

Mulch Layers

A mechanical device that rolls out plastic mulch over a raised bed and with small side plows throws soil over the edges of the plastic to seal it from being blown away by the wind. From there, holes are popped into the plastic and vegetables or strawberries are planted through it. Many times, drip tubing is laid underneath the plastic at the same time. The mulch is used for weed control.

Land Levels/Box Scrapers

There are box scrapers and complex land-leveling implements that are tractor drawn and used for leveling farming fields so that you have a level surface to accurately seed multiple rows of crops and have an efficient, evenly flowing irrigation program. These are necessary for flooding orchards and furrow irrigation of vegetable crops, like corn, tomatoes, lettuces.

Manure Spreaders

If you have livestock and accumulate large amounts of animal manure, a manure spreader is a must. It's basically a pick-up bed that has an escalating bed driven by its wheels, and as it moves it rolls the manure out of the back of the bed. Many spreaders can be adjusted to regulate the amount deposited. It also can be used for shredded straw, pine needles, compost, sand, topsoil, and so on. Very valuable in top-dressing vegetable fields and pastures.

Sprayers

From hand-held to backpack sprayers, even in organic operations, spraying for pest management is part of the program. The most efficient is the small battery operating sprayers that attach to ATVs; these 10–30-gallon sprayers work great for spraying weeds or controlling insect pests. They also are good for keeping water on site for weed mower operations and for fire suppression during the fire season.

PREPARING THAT FIRST PIECE OF GROUND

Taking a piece of fallow land and turning it into a field to grow crops is a daunting task. It takes time and patience. Expect to take 2 to 3 years working the soil before you have a great growing area. It will take a couple of years to build your soils and learn what you can or can't grow on that piece of ground. Begin with a soil and water (water source) sample to see what you are beginning with and get some input from your agriculture extension advisor and your local agricultural commissioner's office staff.

Land Preparation

Let's start at ground zero. We'll take a piece of fallow land, an empty lot, or an area cleared of trees or vegetation, and walk through the steps to bring fields into production.

To start, there are two things to remember. One, Sierra topsoils are usually shallow, acidic, and have low fertility, and two, anything you disturb will open up a weed seedbed that is going to explode. In most cases I recommend that you plan on 2 full years of working that soil before you put it into production.

During the first season, remove the existing vegetation and identify the species that you are removing. Are they forest trees, for which you will have to remove stumps and roots, or manzanita and buckbrush? Is it a vacant lot that that had been disturbed before? Determine what you know about its previous uses, and whether you need to test the soil for substances that may have been in contact with that piece of ground like oils, pesticides, or metals.

For this purpose I'm going to select a piece of forest ground that is very common in the Sierra at about 4,000–5,000 feet elevation. This site has a clearing within a forest community, composed of sagebrush, bitterbrush, rabbit brush, and a few serviceberry or chokecherry. Remember forest topsoils may only be 12–16 inches deep, and in some areas less than a foot. So the key is not to bring the subsoil to the surface. The first step is to remove the larger brush. If your

Ground zero, or your empty canvas.

soils are shallow, you may want to backhoe just the individual species and not turn all the ground over, for fear of disturbing the fragile topsoil. If you have a solid brush land community, you want to start early in the spring and remove the top vegetation. There are many ways to do this; you can do it by fire (coordinating a burn with your local California Department of Forestry or fire district), by animals, whereby you bring in browsing goats to eat the top growth; or by mechanical removal, by mowing or using a bulldozer to blade out the vegetation. If you use a dozer, be careful not to scrape off the topsoil as it removes the vegetation. Then, stockpile the brush and burn it, chip it, or have it taken to a green waste facility. The idea is to disturb the brush in its active growing stage, to stress it so that it will die during that first year. If you do this process in the fall as it's going dormant, the food reserves in the roots will just wait for spring to shoot up again. All the existing top vegetation of the brush should be mulched with a brush hog or flail mower so that it is level to the ground and all that is left is the root stock.

The next step is to turn over the roots to let them air dry throughout the summer. Take an earthen plow and turn over the roots and let them dry for the summer. If you have a ripper or chisel, you can run that 12–18 inches behind the plowing and it will assist in pulling out the roots. All large roots over 1-inch diameter should be physically removed by hand or with a spring harrow.

As you enter late summer and early fall, try to time a series of disking (turning over of the ground) after a late summer rain to begin to break down the root structure of the old shrubs or the sod of the bunch grasses that preexisted. If you have a rocky field, this is the time to hand-pick rocks or run a rock-picking implement through the fields to remove rocks and debris. Continue to cross-disk the field several times to break down clots and incorporate the existing vegetation. It is best to rip the field with a field chisel or shank at least once that fall to allow winter rains and snow to percolate nutrients within the soil levels. If the field had no shrubs and was just a fallow pasture, this is the time that you can then add compost or manures and establish a cover crop of rye, legumes, and clover. Run a toothed harrow or screen (cyclone fencing) to get a rough crop of vegetation started to hold your soil through the winter.

In the spring, monitor what germinates from this new seedbed. Allow all the species to grow from the field, weeds and all, and identify as much of the species as you can. By late spring–early summer, as the ground dries out and before the vegetation turns to seed, you want to disk under the vegetation at least twice, cross-disking in either direction to kill the weeds and incorporate the green manure. If possible, add at least 1 inch of water once a month to stimulate weed seeds to grow, and then disk them under to help reduce your weed seed bank in that field. Throughout the summer continue to remove

Turn over the roots so they can dry out in preparation for disking.

rocks, roots, and debris to help improve the structure of your soil. It's also a great place to disperse fresh manure from your animals, or composting materials like leaf cuttings, straw, and so on. The idea is to work and improve the structure of your soil throughout the second year, adding organic matter and incorporating green manures along the way. As you get to fall in year 2, take a final soil test of your field to check pH, organic matter, NPK (nitrogen, phosphorous, potassium), and your secondary nutrients of calcium and magnesium, along with any salinity or sodium deposits. These tests should be done every 3 years from here on. Your soils will tend to be slightly acidic; if they are below 5.5 pH and you are planning to grow vegetables, you may have to consider adding lime and large amounts of vegetable compost, worm castings, and straw. Your goal for vegetables is a pH between 6.0 and 7.5. This is also a great time to look at the structure of your soil, with the canning jar test, to separate out your sand, silts, and clays and establish what your loams will be, for example, sandy-clay loam or silty-clay loam (see page 21). Finish in the fall with a nice cover crop that best suits your area and you will be ready to seed and plant your crops the next spring. The key is patience and planning. Don't try to rush a field into production. I even allow three seasons for mine so that the soil structure and pH have had time to stabilize and to minimize my weed seed banks. After the second year is also a great time to add in your permanent irrigation systems and your access roads, utilities, and so on.

One word of caution: If you are in an area with oaks, have removed some, and plan to put in an orchard of fruit trees, beware of oak root fungus and phytophora disease to fruit trees caused by oak roots that have been left behind. Consult your local agricultural commissioner's office to see if there have been any previous problems.

Seedbed Preparation

In the Sierra, late spring is the time to begin preparations of all of your fields. This system is what I use to prepare all 6 acres of my organic vegetable fields. I first keep an eye on the fields after the snowmelt, usually around the first of April. After the first couple of weeks of drying winds and full sunshine, I drive a shovel in the soil to see how wet the soil conditions are. If the soil sticks firmly to the shovel it's too wet, but if it falls off and crumbles evenly through my fingers it's ready "to be opened up." Opening up means to run a disk over it to open the soil and allow air and warmth into the soil so that it can dry further and warm up. Over the winter, heavy rains and snow pack will compact your soils, and you first have to open them up and let air and warm temperatures enter the soil. This should be done 1 or 2 weeks before planting. Once the soil is opened up, you can run a ripper or chisel through it to break up the subsoil for better root growth, especially if you are planting root crops like potatoes, carrots, turnips, or beets.

Moisture is the key to the preparation of a seedbed. You want just enough soil moisture to cultivate the soil in a nice loose manner, but if it's too wet and clumps, or compacts under the tractor tires you can ruin the whole structure of your field; too dry and it will not crumble the dirt clods, and prohibit water absorption and proper seedling establishment. The soil should have 50–75 percent field capacity of moisture, which is a slightly dark color, moist to the touch that crumbles and slightly balls up, or makes a broken ribbon when squeezed. Next, using your soil thermometer, once the soil reaches 50–55 degrees within the top 2 inches you're ready to seed or put in your transplants.

My 6 acres are always disked the first time in the spring to open the soil and kill the spring growth of weed seedlings. Next, plot out the size of the field you plan to plant. We plant a succession of direct seedlings every 2–3 weeks: lettuces, spinach, chard, arugula, radishes, carrots, beets, and mizunas. We start with about a quarter of an acre around the end of April and continue to seed quarter-acre plots until September 1.

The preparation of the seedbed consists of a single disking, followed by a shallow 3–4-inch depth of rototilling, then rolled with a ring roller to set the seedbed. It is very important to use a water roller or a ring roller to set the seedbed after rototilling so that the seeds are not planted too deep. Once the rolling is done you are ready to seed or transplant.

For direct seeding, there are many different ways farmers seed their crops. Some like to use a bed shaper, which is an implement pulled behind a tractor that shapes the soil into 40-inch raised beds, up around 6 inches in height, which are then seeded by a variety of seeders. This is also used to transplant seedlings with a mechanical transplanter. If you are planning to use a plastic mulch layer, and transplant your seedlings through the plastic mulch, this is the time you would do it.

I like to manually direct-seed my fields of lettuces, radishes, carrots, beets, arugula, mizunas, broccoli raab, spinach, kale, and Swiss chard with a Planet Jr. walk-behind seeder. You can pull a string to seed straight rows, or you can be like me and eyeball it, walking a relatively straight line, pushing your seeder. I call it "feng shui farming": the rows may be a little crooked, but they still grow just as well as if they were in a straight line, maybe even better. I don't use a bunch of seeders on a tool bar behind a tractor, because I try and minimize compacting my soil more than I have to. My grandfather and my dad used the same walk-behind Planet Jr. drop seeder to seed 5–10 acres of row crops of cut flowers in San Jose, and it worked fine for that amount of acreage—you just did a lot of walking. Hey, exercise is good!

By the second year, your soil will be greatly improved. Here is a disked field.

Once the seedbed is disked and rolled you are ready to seed or transplant.

The key to seedbed preparation is the timing of turning the soil over with just the right amount of moisture. In the spring, I usually have plenty of moisture in the soil. Because I continue my succession plantings of cool-season crops throughout the summer, soil moisture decreases rapidly in July and August, even into September, and I have to presoak the field that I'm planting on so that I have enough moisture to prepare the seedbed. Too many times I see farmers pulverizing fields leaving a cloud of dust, because they're as "dry as a popcorn fart," and there goes your soil in a windstorm. With just a little moisture and a rough disk, instead of a fine till, you will save your soil and have it ready to seed when you need to. Always try and prepare your soil the day of seeding so that you have fresh moisture available for the seed and to help the capillary action of seeds absorbing your irrigation water. It's very important that you irrigate soon after seeding. If you prepare the soil and wait to seed a few days later, you will lose that moisture, soil will tend to repel water, and you will have spotty germination because of the inconsistency of water being absorbed by the seed.

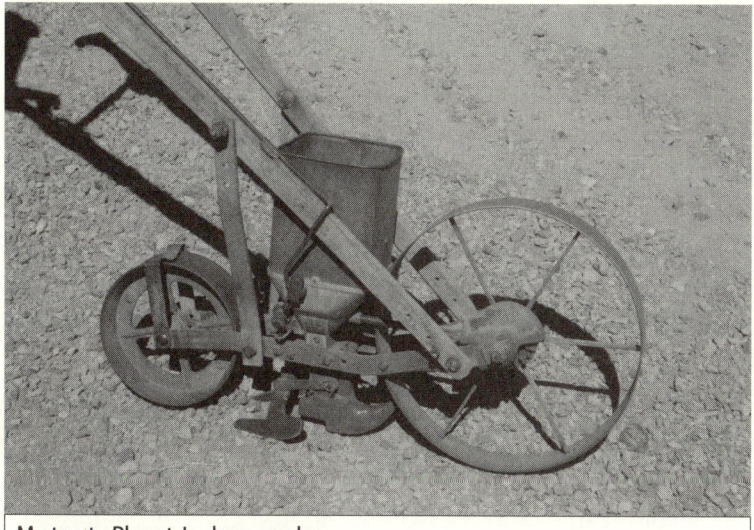
My trusty Planet Jr. drop seeder.

Cool-Season Vegetable Production in Succession
Successive plantings of cool-season crops makes successful farming possible in the Sierra above 4,000 feet. Over the past 15 years, here at Sierra Valley Farms we have been farming cool-season crops from the end of April until the last harvests at the end of October outdoors with no row covers. Yes, it's possible to have a 6-month growing season in the Sierra. Frost and all. The

reason for direct-seeding all cool-season crops (arugula, lettuces, mizuna, carrots, radishes, beets, spinach, and kales) other than broccoli, cabbage, kohlrabi, Brussels sprouts, and herbs, is that it allows them to adapt to Mother Sierra's climates and microclimates. Transplants of lettuces in the Sierra are very difficult because they are subjected to wind, sun, rain, and even hail. During the early summer and into fall, there can be as much as a 50–60 degree temperature change from morning to evening. It's very stressful for lettuce transplants, and most don't make it or tend to bolt because of the stress. Plus, it is best to grow leaf lettuces in the Sierra versus head lettuces. Head lettuces, like iceberg, and the "ball" lettuces will not make a dense head because our days of the season are too short and they don't develop right. When direct-seeding, the young seedlings adapt to the daily temperature fluctuations and develop thicker cell walls to cope with the wind, rain, sun exposure, and heat. In case of a frost, you may lose that one cutting of leaf lettuce, and if you cut that back, within 5 days you will have another crop. For head lettuce, once the main head is damaged your crop is ruined.

After the soil is prepared for the seedbed we direct-seed one-fifth of an acre of the same crops (lettuce mixes, arugula, kale, spinach, romaine, radishes, carrots, beets, and Swiss chard) every 2–3 weeks, with 7–8 plantings a summer. Our last planting is usually around Labor Day. This successive planting is for loose-leaf greens production, not for head lettuce production. In the Sierra, you can easily get three to four cuttings of leaf lettuces compared to one cut of a head lettuce. You can increase your income per linear foot three times as much for leaf lettuces as you can for individual head. As for root crops, radishes, carrots, turnips, parsnips, beets, we direct-seed them and do not thin. We harvest the outer leaves when they become ready; that thins the crop, and the next ones will fill in. My philosophy is to sell everything that I plant, and weed everything I don't plant. As for my specialty cool-season crops like broccoli, broccoli raab, cauliflower, and cabbage, I start them twice (3 weeks apart) in the greenhouse in plug trays in late March and mid-April, harden them off in a cold frame, and transplant the first crop by Memorial Day and the second by June 15.

I've developed this model on the assumption of "What I could maintain by myself if I didn't have any help?" I came up with about one-fifth to a quarter of an acre, for which I can prepare the soil, seed, weed, irrigate, and harvest by myself. If I get some extra help then that's even better. I will outline my operation for you. Realize that I am able to do this because of the equipment I have and the fact that I was born with a rubber spine, so I'm able to stay stooped for long periods of time. So to be safe, a family of three or four can easily do this model.

I begin with a plot that's roughly 70 feet by 120 feet. The plot in April is first disked to open the soil, allowed to dry for a few days, then cross disked, lightly rototilled to a depth of about 4–5 inches, then rolled with a ring roller. The seedbed is now prepared.

Next I lay out my manifold of one-half-inch drip tubing, in which I have 20–25 T-Tape barbed-adapters punched in about every 18 inches to 2 feet. T-Tape ranges from 6 mil to 15 mil plastic tubing that has slits in the tubing every 6–8 inches that allows a small bubble of water to trickle out. It runs on low water pressure (10 psi), and a garden hose can pretty much water 25 rows, 100 feet long. It comes in a roll ranging from 500 feet to over 6,000 feet and can be used for 2–3 years.

I direct-seed my crops into flat ground and not-formed beds with my Planet Jr. It is important to keep your soil temperatures as cool as possible in the summer; not elevating the soil keeps your temperature down. Raised beds will absorb the daily warm temperatures and make your lettuces bolt and root crops like radishes go to seed.

Irrigating seed with T-Tape in a hoop house.

Each one of these drip manifolds is attached to a garden spigot. I don't use a pressure reducer because I like to carry out as much water as I can in the longest manifold that I can; friction loss in the drip tubing over a long run will be just as effective as a pressure reducer. You can also adjust your hose bib to control your water pressure. If you are watering from a well, reservoir, or ditch water, you must have a wye filter to remove sediment deposits or they will clog your emitters.

Once your manifold is laid out, put a landscape stake at each barbed fitting to hold it in place. Next is seeding. Whatever seeder you are using, and whatever vegetable you're seeding: *Never plant a seed deeper than three times its width of that seed*. I plant all my seed cool-season crops no deeper than one-quarter of an inch deep, pretty much just under the surface. For the Planet Jr. drop seeder I like to use a whole two times the length of the seed so that seed will flow smoothly out as I push the seeder ahead of me. I like a thick row of leaf lettuce: one, it gives you more dollars per lineal feet, and, two, it helps crowd out weeds so that you won't have to do more hand-weeding during harvest. Same goes for carrots, radishes and beets; I don't thin, I leave them to crowd each other to prevent weeds, then as the outer ones develop I pick those and the inner ones develop for the next week. This gives me 3–4 weeks of harvesting a row with a smaller-size root that specialty chefs like and will pay more for, and it's all about increasing the dollar on your crop. A few 2-pound beets in one row are not going to pay the bills, but a few thousand small ones will.

If you like straight rows, now is the time to pull a string from each barbed fitting to the end of the row about 100 feet long. As I said earlier, I eyeball my rows, pushing my Planet Jr. as I go, seeding in a somewhat straight manner. When I reach the end of the row I turn around move the seeder over 4–6 inches, then seed down the other side. I continue this pattern at each barbed fitting across my manifold. For my crops, I divide my 20–25 rows by the crops that I'm growing, for example, lettuce mixes, arugula, romaine, spinach, Swiss chard, kale, carrots, and beets, usually planting 2–4 rows of each depending on what I think I can sell. Make sure you label each. I log each one of my fields, rows, and crops.

In this production scheme, each crop allows me to sell at four farmers markets a week, supply a CSA and four restaurants. Add that up for a steady stream of income for 6 months and that's a lot of produce! Each crop will give you about 4–5 weeks of steady production. The lettuces and radishes are first to produce in about 40 days, and the first to burn out in about 4 weeks. I continue to water the greens and allow them to flower to bring in my beneficial insects, and then to seed, to select my hardiest varieties to collect. You can install barbed shut-off valves to then shut them off, or as I do, just cut off a 1-foot section, tie it in a knot, and place it into the barbed fitting as a "plug." The carrots and beets will take longer, about 70 days before harvest. You then move to the next successive crop after about 5–6 weeks.

Once you're done seeding the rows, it's time to pull the T-Tape. I take the roll and place it between two homemade sawhorses and walk down each row, pulling and laying the tape between the two seeded rows of each barbed fitting. At the end I tie the end in a knot, slide a 6-inch steel landscape staple within the knot and secure it into the soil. As I walk back to the roll I place a landscape stake about every 30 feet to secure the T-Tape in the center of both rows. This prevent the Sierra winds from blowing it off the rows. When I get back to the roll I cut it off and attach it to the barbed compression fitting. That row is now complete.

You continue to install all 20–25 rows in that manner, until you have completed your manifold. Once you've completed all the T-Tape installation you can turn on your water faucet and charge the lines. It may take 15–20 minutes to charge. While it's charging, walk each row and make sure the T-Tape is between the two seeded rows so that you have even percolation water to each row. As the pressure builds, squeeze the T-Tape to see how much pressure it has; you don't want it too tight or it can bust, you want just a little squeeze (don't we all!). You are now in operation.

The idea of the T-Tape is to put water only where you want it: on the seed. Wherever you put water weeds grow, so you want to minimize over watering. T-Tape will sweat a band of moisture between the two seeded rows, which in this first application can take anywhere from 45 minutes to over an hour. When the band of water has reached across both rows you now shut off the water.

You want to keep the soil moist over the next few days, then once it dries water again. It will take less water each time. Your crops will germinate in 5–10 days depending on the species. Radishes and arugula are usually the first, with lettuces, spinach, and kale to follow. Carrots and beets are the last to germinate, which can take 10–14 days.

Next comes the fun part: weeding. The key to a successful crop is getting 80–90 percent of your weeds out of the crop in the first 2–3 weeks. If it gets away from you here, it's pretty much all over; you might as well disk it under for a cover crop.

The good news is that arugula, lettuces, and radishes are the first to emerge within a week, then spinach. Carrots and beets are the last and the most crucial. You want the greens to get about a week old after emerging before weeding so that their roots have taken hold. I exclusively use hula hoes and first go down the outside of the rows lightly and then pull the T-Tape from the middle and weed the center. You will then have to hand-pull the ones that you can't get with the hula hoes. In between each row you have a walking aisle between 16 and 18 inches that will develop weeds. It's your choice, you can either run a small tiller on the small weeds, or a wheel hoe, or you can just walk on the weeds. I tend to let them grow; as they mature, flower, and seed they provide a fall cover crop for me, attract beneficial insects, and provide seed for birds. Once every 3 years I sprinkle a cover crop mix between the rows, and after harvest add overhead sprinklers to develop a nice fall cover crop and mulch/ disk it in for winter.

My weed program concludes with a third weeding just 2–3 days before harvest. We try to hand-weed all the crops that we are going to harvest the upcoming week, so that they are nice and clean, and we don't have to weed as we harvest. It sounds good, but that doesn't always happen, so weeding is an ongoing project here on the farm. It's just part of farming. Once the crop is done harvesting we move to the new crop, and all maintenance seizes on the older crops.

This is our system, and our crops are rotated according to the field they were grown in the year before. It's best to label your fields (use numbers or names) and then rotate your fields at least every 2–3 years, so that you are not planting the same crops in the same place every year. Broccoli, broccoli raab, kales, and cabbages replace our rotation of lettuces, spinach, radishes, carrots, and beets. All of our new fields first begin with garlic, green onions, leeks, and shallots because we can seed them in the fall and dry land farm them for a crop in June–July. Then the following season we begin our rotation of cool-season crops.

Season's End

Mountain farming, like any other type of farming is hard work, a little income, and a lot of love—you've got to love it! Establishing a farm is an ongoing project, long hours, frustrations, failures, and successes. Yet the beauty of farming is that it's your masterpiece! It represents who you are, your beliefs, and the way you want to live your life.

We conclude our outdoor season at the end of October and into early November with all the lettuce-type crops, the T-Tape is all pulled from the fields, and all the irrigation is shut off and winterized (by draining the water from the pipes). We continue to harvest carrots, beet, horseradish, and radishes until the ground begins to freeze in early December. Before the ground gets too wet, we mulch mow all the fields down to the ground and turn the ground over once with the disk. The crops have been put to bed for the winter, and we anxiously wait for spring and start all over again.

In summary, there is no more rewarding job than being a farmer. The pleasure of working outdoors surrounded by Mother Sierra's beauty and wildlife, growing food for my family and local communities, and being at peace with myself, where nothing else really matters but family and health, and leaving a legacy for my children. It just doesn't get any better!

Sierra Valley Farms

REFERENCES

Coleman, Eliot. *The Winter Harvest Handbook*. White River Junction, VT: Chelsea Green Publishing, 2006.

Folchi, Betty. *Notes of a High Country Gardener*. Cottage Grove, OR: Morning Dove Press, 1995.

Gleason, Chris. *Building Projects for Backyard Farmers and Home Gardeners: A Guide to 21 Handmade Structures for Homegrown Harvests*. East Petersburg, PA: Fox Chapel Publishing, 2012.

Hill, Lewis. *Cold Climate Gardening: How to Extend Your Growing Season by at Least 30 Days*. Pownal, VT: Storey Communications Inc., 1987.

Nones, Raymond. *Raised-Bed Vegetable Gardening Made Simple*, 2nd Ed.: Woodstock, VT: Countryman Press, 2013.

Philbrick, Helen and Gregg, Richard. *Companion Plants and How to Use Them*, 3rd Ed.: Devon-Adair Company, 1976.

Romano, Gary. *A Hiker's Guide to Wild Edible Plants of San Luis Obispo County*. Menlo Park, CA: El Camino Publishing, 1982.

Storer, Tracy I. and Usinger, Robert L. 1963. *Sierra Nevada Natural History*. Berkeley, CA: University of California Press, 1963.

Swain, Ralph B. *The Insect Guide: Orders and Major Families of North American Insects*. New York: Doubleday, 1948.

United States Department of Agriculture. *Soil Survey of Sierra Valley Area, California, Parts of Sierra, Plumas, and Lassen Counties*. University of California Experiment Station, 1967.

Whitney, Stephen. *A Sierra Club Naturalist's Guide*. San Francisco, CA: Sierra Club Books, 1979.

Young, Carol L. *What Grows Here? Gardening in Plumas, Lassen, and Sierra Counties*. Chico, CA: North Valley Printing, 1986.

ACKNOWLEDGMENTS

There are many people to thank for this book and my previous book, *Why I Farm: Risking It All for a Life on the Land.* It's a lifelong laundry list of those who have influenced me, helped me at the crossroads, and steered me on the path to who I am today. The thanks begin at home, with my parents, Rose and Louis Romano; Grandpa Giovanni L. Romano, who taught me that I would be rewarded for hard work and to treat people as I would like to be treated; and, finally, to my twin brother Larry, my mentor. Larry, I really miss you, you definitely died too young, and I would have enjoyed working with you on this book. I'm sure I left out a lot of good stories, because you never forgot anything, unlike me, who would lose my head if it weren't attached to my body.

Next, I acknowledge my inspirational leaders, Uncles Raymond, Emilio, Benny, Beno, and Marion Folchi. I bought the ranch from Uncle Emilio and Aunt Betty 26 years ago—they were all great storytellers and honest, hardworking people, many of whom also died too young.

Along the way came a core of friends that I call "The Most Interesting Friends in the World," a menagerie of personalities who have always been there for me. Even after months or years apart, we can get together and pick up where we left off, and 35 years later we're still on the same page: Uncle Bob Belton (Reno . . . Reno), Terry Goll (Let's Party!), Tony Piazza (Chic), Dan Gianini (Fishbait), Dave Dickinison (Skate), Mark Ludlow (Ludlow), Bob Bolton (Scottsdale), and Roland Haga (Skoll), and many others.

I also give credit where credit is due. To my first wife, Tami: We brought into this world the best daughter that a father could ask for, Elizabeth. You have always been the best child and a wonderful person, excelling in everything you've accomplished, and I am so proud of you! Tami was instrumental in helping me set up the nursery when we first started, and we were able to overcome our differences and remain friends today. Thank you for being there during the early times. My second wife, Kim: Thank you for hanging in there with me for all those years, helping to make Sierra Valley Farms what is is today. I could not have done it without you. Thank you, too, for the most incredible son a father could ever have, Joey. It's a mystery where he got the talent in sports, but watching him play baseball is a joy for this father, who never was able to play sports because I had chores on the farm. I live my sports life through you, Joey. Keep up the great work and go to Cal Lutheran and on to major league baseball—we're all rooting for you! If Joey Romano isn't a major league baseball name, I don't know what is. Love ya, Joe!

My first book was a project that had been in my mind for a few years, and thanks to a light winter in 2012—not having to plow and shovel snow for 5 hours a day—I was able to sit down at the keyboard and go crazy. I thank Mayumi Delgado from Moonshine Ink for her direction on publishing *Why I Farm* with Bona Fide Books in South Lake Tahoe, Lis Korb for editing and helping me to structure the outline, Roger Freeberg for his wonderful photos, and Lauren Shearer for the graphics.

I thank the reviewers who took the time to review my manuscript and made fabulous contributions to make *July & Winter: Growing Food in the Sierra* as comprehensive as possible. These people include SunMie Won, Eric Larusson and the staff at the Villager Nursery in Truckee, Sally and Frank Massimino of Green Cedar Farms, Carolyn Meiers of South Lake Tahoe, and Michelle McLean of Meyers, California.

Finally, my gratitude goes out to Kim Wyatt of Bona Fide Books, who took a chance on this dirt farmer, and still does. She believes in my projects and the message that I am trying to get out to the general public that we need to grow our own food as small farmers, home gardeners and building our local food sheds. Thanks for your patience and direction. I look forward to working with you on my next book, yet another attempt to promote organic farming, farm living, and a healthier America.

After a childhood spent pulling weeds and planting seeds, Gary Romano received a master's degree in recreation administration from California State University, Chico, and worked as a park ranger and county park administrator before returning to the farm. To find out more about Gary and his farm, go to www.sierravalleyfarms.com.

INDEX

Note: Pages referring to figures are in italics. Tables are indicated with a t.

A

Acidity, 22, 23, 31
Acorns, 90
Alkalinity, 22, 23
All-terrain vehicles (ATVs), 176, 179, 181
Almonds, 90
Alpine region, 4
Animal husbandry, 143. See also Chickens
Annuals, 19, 53, 55
Aphids, 135, 140
Apples, 76
    suggested varieties, 76–78t
Apricots, 78
    suggested varieties, 79t
Artichokes, 60–61, 154
Arugula
    direct seeding, 40, 42, 55, 186, 190, 192
    germination, 56, 193
    growing outdoors, 148
    harvesting, 58
    in heated greenhouses, 152
    in hoop houses, 13
    identifying, 61
    insect pests, 43, 137
    outdoors, 14
    succession planting/rotation, 57, 124, 186
Asparagus, 60, 150, 157
Axes, 50

B

Backhoes, 178, *178*
Basil, 39, 57, 125, 152. See also Herbs
Beans
    bush vs. pole type, 61–62
    direct-seeding, 186
    growing, 61–62
    in hoop houses, 39, 152
    insect pests, 137, 138
    planting outdoors, 43, 107, 190
    on trellises, 36, 46
    when to plant, 56, 57
Bears, 134–135
Bed shapers, 180
Beets
    direct seeding, 42, 55, 190, 192–194
    growing in hoop houses, 126
    growing outdoors, 56, 65, 148
    insect pests, 136, 137
    nutritional requirements of, 24
    successive plantings, 57
    watering, 126
    when to plant, 56
Berries, 39–40, 46, 73, 93–94, 168, 171
    blueberries/huckleberries, 22, 31, 94–95
    brambles (raspberries and blackberries), 95–97, 102
    cranberries/lingonberries, 31, 95
    currants and gooseberries, 96
    grapes, 99–100
    how to grow, 94
    logan and boysenberry, 100
    protection from wildlife, 94, 133, 143
    strawberries, 101–102
    watering, 46
    See also Native berries and fruits, 102–105
Biennials, 53
Biodiversity, 9, 16–17, 18, 19–20, 143–144
Biomass, 31
Birds, 133
Blackberries, 96–97, 102
Blade mowers, 178–179
Blades, 178
Blister beetles, 136
Blood meal, 30, 130
Blowers, 179
Blueberries, 22, 31, 94–95
Bone meal, 30
Box scrapers, 180
Boysenberries, 100
Brassicas, 11, 33, 39. See also Broccoli; Cabbage; Cauliflower
Broccoli
    in crop rotation, 194

203

germination, 57
growing in greenhouses, 156
growing in hoop houses, 152
growing outdoors, 35, 60–61
harvesting, 61
nutrient use of, 24
outdoor succession planting, 148
rotating, 33
transplanting, 41, 55, 57, 122, 190
watering, 44, 61
See also Brassicas
Broccoli raab, 41, 61, 71, 156, 186, 190, 194
Brussels sprouts, 24, 41, 44, 55, 57, 60, 61, 191
BT (Bacillus thuringiensis), 139
Buckets, 178, 178
Budgets
   capital, 167–168
   operating, 168–169

C

Cabbage
   growing in greenhouses, 156
   growing in hoop houses, 152
   growing outdoors, 148
   harvesting, 58
   insect pests, 136–137
   nutritional requirements of, 24
   rotating crop, 33
   transplanting, 41, 55, 57
   watering, 44
   when to plant, 56
Cabbage loopers, 136, 139
Cabbage worms, 139
Calcium, 24
California blackberry, 102
California map, xviii
California Native Plant Society, 9
Cantaloupe, 63
Capsaicin, 130–131
Caraba bugs, 138
Carrots
   direct seeding, 40, 42, 55, 190, 192
   germination, 56, 193
   growing and harvesting, 65, 66
   growing in hoop houses, 13, 15, 148
   growing outside, 14, 65, 126, 186

nutritional requirements of, 24
rotating, 194
successive plantings of, 57, 116, 186
Cation exchange (CEC), 22
Cauliflower
   growing and harvesting, 60, 61
   growing in greenhouses, 156, 190
   growing in hoop houses, 152
   growing outdoors, 61, 148
   insect pests, 136, 137, 139
   nutritional requirements of, 24
   transplanting, 41, 55, 122
   watering, 44
   when to plant, 56
See also Brassicas
Celery, 24, 41, 55, 57, 154
Chestnuts, 91
   suggested varieties, 91t
Chickens, 27, 38, 87, 129, 136, 138
Chipmunks, 130–131
Chisels, 177–178
Chives, 57, 109, 111, 116, 134, 154
Chlorosis, 22
Chokecherries, 103, 183
Cilantro, 125, 135, 148
Clay loam, 21
Cleat tractors, 173, 177. See also Tractors
Climate, 5–6
   microclimates, 7–8, 9, 17, 50, 73, 190
      See also Frost; Precipitation; Temperature; Weather
Club root, 33
Cold frames, 37, 38–39, 107, 109–111, 110. See also Hoop houses; Season extension structures
Coleman, Eliot, 62, 109, 121
Colorado potato beetles, 136
Community-supported agriculture (CSA), xvi, 168, 169, 172, 192
Companion planting, 129, 158–159t
Composting
   with cover crops, 31
   of manure, 27–28, 38
   structures for, 26–27
   three-box system, 26
Composts
   acidifying, 22
   activators for, 25

adding to crops, 28–29
organic, 24–25
Cool-season crops, 11, 13, 14, 15, 35, 39, 40–41, 53, 190–191
   direct-seeding, 57
   in season extension structures, 124–125
   watering, 46
   See also individual crops by name
Corn, 11, 24, 35, 36, 41, 43, 46, 55, 56, 57, 61, 62, 107
Cover crops, 15, 31, 33, 194
   for fruit trees, 87
Cranberries, 31, 95
Crawlers, 173. See also Tractors
Crop maintenance, machinery for, 171
Crops
   hybrid varieties, 54
   flowering, 60–61
   planning, 124–125
   planting, 41–42
   rotating, 23, 33, 119, 120, 124, 136, 140, 168, 194
   selection of, 39–40, 53–54
   to transplant, 40–41, 54–55, 122–123
   warm-season, 11, 35, 39, 41, 57
   See also Cool-season crops; Direct seeding; Transplants
Cross-pollination, 69
Cucumber beetles, 136
Cucumbers, 14, 41, 46, 57, 61, 62, 125, 136, 152
Cultivators, 49, 177
Currants, 96, 98–99
   suggested varieties, 98t
   wild, 104

D

Deer, 65, 133–134
Diggers, 178
Direct seeding, 40–42, 54–55, 186, 190, 192–194
   of cool-season crops, 57
   on the east slope, 41
   in season extension structures, 121
Diseases, 142
   club root, 33
   oak root fungus, 185
   phytophora disease, 185
   scab, 33
   verticillium root rot, 93
Disks, 160, 176, 176, 188
Drip irrigation, 42–43, 171, 191–192
   for fruit trees, 84, 85

E

Earthworms, 25–26
Earwigs, 136
East slope
   climate, 5–7, 13–14
   crop selection, 39
   direct seeding, 41
   exposure, 5
   fall frost, 15
   fruit trees, 87
   hardiness zones, 10
   precipitation, 3
   prickly pear cactus, 89
   site selection, 36
   soil, 22
   summer crop planting, 56
   See also Sierra region
Eggplant, 24, 39, 41, 43, 46, 55–57, 61, 62–63, 93, 107, 122
Elderberry, 102–103
Electrical conductivity, 22
Elevation, 6, 9, 11
Equipment. See Machinery and equipment
Erosion, 33
Exposure to slope, 4–5, 9, 11, 35

F

Farmer's Almanac, 56
Farming
   as business, 161, 163
   crop plan/profit and loss, 168–169
   equipment and supplies, 168, 172–181
   facilities, 166–167
   farm management, 169–170
   financials and budgets, 167–168
   labor, 168
   land preparation, 181–185
   machinery and equipment, 170–181
   miscellaneous expenses, 169
   planning, 163–164

sales and marketing, 172
season's end, 194–195
seedbed preparation, 41–42, 185–189
water, 164–166
Fertility, in season extension structures, 120
Fertilizer boxes, 179
Fertilizer spreaders, 49–51
Fertilizers, 29–31
Figs, 88
    suggested varieties, 88t
Flail mowers, 179
Flaming weeds, 141
Flea beetles, 137
Flora and plant communities, 3–4, 9. See also Native flora
Flowering crops, 60
    artichoke, 60–61
    asparagus, 60
    broccoli, Brussels sprouts, cauliflower, 61
Flowers
    annuals, 19, 53, 55
    bulb crops, 28
    companion planting with, 129, 133, 135, 139–140
    creating wildlife habitats with, 19, 143
    direct seeding, 55
    fertilizing, 29, 30–31
    growing, 53, 116, 120
    perennial, 19, 29, 53, 125, 150
    quince, 90
    saving seeds, 69, 71, 72
    sunflowers, 46, 55, 92–93, 133, 149
    transplanting, 55, 107
    watering, 46
    when to plant, 56–57
    wildflowers, 150
    See also Flowering crops; Fruit trees; Native flora
Foothill region, xvi–xviii, 1. See also West slope
Forks, 48
Frost, 10–11, 15
    frost-free days, xvii, 9, 10–11, 11t
    protection from, 123–124
Fruit trees, 39–40, 73
    apples, 76–78

apricots, 78
care of, 84–87
cold hardy, 75
harvesting and storage, 75–76
heirloom, 75
peaches, 78–80
pears, 80–81
planting, 82–84, 83
plums, 80, 82
protection from wildlife, 86, 87
pruning, 84–85
site selection, 73–74
specialty fruits, 87–90
spraying, 86, 171
variety selection, 74–75
watering, 46
whitewashing trunks, 84, 87
See also Native berries and fruits
Fruits
grapes, 99–100, 104
melons, 63
See also Berries; Fruit trees
Fungus, 142
Fungus gnats, 140
Furrow irrigation, 43, 122

G

Gardening tools, 47–51t. See also Tools
Garlic spray, 137
Garlics, 24, 28, 56, 65, 67, 109, 111, 134, 136, 150, 154, 176
Geodesic domes, 107, 114, 116. See also Season extension structures
Germination, and soil temperature, 56
GMO crops and seed, 53, 69
Gooseberries, 96, 98–99
    suggested varieties, 99t
    wild, 104
Gophers, 130–132
Gourds, 63, 69
Grapes, 99–100
    suggested varieties, 100t
    wild, 104
Grass sod crops, 31
Green manure, 31, 33, 185
Greenhouse planting schedule, 156–157
Greenhouses, 14, 57, 107, 112, 115–116.

See also Hoop houses; Season extension structures
Greens, 41, 42, 50, 55, 56, 57–58, 59, 120, 121, 126, 149, 159
    seeding, 121
    See also Lettuces; Microgreens; Mizuna; Mustard Greens
Ground squirrels, 130–131
Growing schedules
    heated greenhouse planting, 156–157t
    hoop house planting, 152–155t
    outdoor planting, 148–151t

H
Hardiness zones, 10–11, 40
Hardy Kiwis, 88–89
Harvesters, 171–172, 179–180
Hawthorns, 89
Heat intensity, 14
Heirloom seeds and varieties, 53–54, 75
Herbicides, 141
Herbs, 151, 152, 157
    basil, 39, 57, 125, 152
    chives, 57, 109, 111, 116, 134, 154
    cilantro, 125, 135, 148
    lavender, 134
    thyme, 134
Hoes, 48, 193–194
Honeydew, 63
Hoop houses, 15, 38, 39, 57, 107, 108, 112
    high hoop houses, 113–114
    insect pests in, 139–140
    irrigation in, 191
    low hoop houses, 111, 113
    planting schedules for, 152–155
    Sierra Valley Farm examples, 112
    site selection, 109
    soil moisture in, 142
    tunnels, 114
    ventilation for, 114–115, 115
    watering in, 46
    See also Cold frames; Season extension structures
Hops, 89
Horseradish, 67, 149
Huckleberries, 94–95

Humus, 25
Hybrid varieties, 54

I
Icing, 73. See also Frost
Insects
    beneficial, 17, 19, 86, 105, 139–140
    indoor, 139–140
    pests, 135, 135–138t, 139
Iron deficiencies, 22
Irrigation, 41, 42–46, 164–165
    machinery for, 171
    T-tape placement for, 44–45, 191
    See also Drip irrigation; Watering

J
Jerusalem artichokes, 28, 39, 65, 67

K
Kale
    direct seeding, 40–41, 42, 55, 186, 190, 192
    germination, 193
    growing and harvesting, 58
    growing in hoop houses, 13, 14, 15, 153
    growing outdoors, 11, 14, 39, 149
    insect pests, 137
    nutritional requirements of, 24
    in raised beds, 35
    rotating, 194
    soil temperature for germination, 56
    successive plantings, 57, 124, 126, 149
    transplanting, 41
Kelp emulsions, 29
Kiwis, hardy, 88–89
Kohlrabi, 24, 41, 55, 56, 57, 65, 151, 156, 190

L
Lacewings, 140
Lady bugs, 140
Land levels, 180
Lavender, 134
Leaf miners, 137
Leafhoppers, 137
Leafy greens. See Greens

Leeks, 57, 109, 111, 148, 194
Legumes, as cover crops, 31
Lettuces
    bolting, 191
    in cold frames, 110
    as cool-season crop, 39, 39
    direct seeding, 40, 42, 50, 55, 120, 121, 186, 190, 192
    end of season, 14, 15, 194
    germination, 56, 193
    growing outdoors, 11, 39, 149
    harvesting, 58
    harvesting seed from, 192
    head, 58, 190
    in hoop houses, 13, 14, 15, 113, 153
    insect pests, 137
    irrigation of, 180
    leaf, 58, 190, 191
    nutritional requirements of, 24
    in raised beds, 191
    sites for, 35
    successive plantings, 57, 116, 149, 190, 194
    transplants, 41, 54, 57, 111, 190
    See also Greens
Lingonberries, 95
Livestock, 143. See also Chickens
Loganberries, 100
Lower montane, 3

M
Mache, 15, 126, 153
Machinery and equipment
    bed shapers, 180
    blowers, 179
    buckets/blades, 178
    chisels or rippers, 177–178
    cultivators, 177
    diggers/trenchers, 178
    disks, 176
    fertilizer boxes, 179
    harvesters, 179–180
    land levels/box scrapers, 180
    manure spreaders, 180–181
    mowers, 178–179
    mulch layers, 180
    plows, 176
    rollers, 177
    selection of, 170–171, 172
    sprayers, 181
    tool bar seeders, 179
    tractors, 172–176
    transplanters, 180
    weed eaters, 179
Magnesium, 24
Manure spreaders, 180–181
Manures, 27–29, 124
    composting, 38
    general make up of, 28t
    as pest repellent, 129
    See also Green manure
Maps
    California, east and west slopes, xviii
    Sierra Valley Farms, xiii
Master Gardener Program, 9
McLean, Michelle, 116
Melons, 63, 151
Mesh tools, 70, 71
Mice, 130–131
Microclimates, 7–8, 9, 17, 50, 190
    and fruit trees, 73
Microgreens, 156
Micronutrients, 22, 24, 29
Microsprinklers, 41, 43, 171
Mildews, 142
Mizuna, 14, 15, 40, 42, 50, 55, 57, 58, 124, 149, 153
Molds, 142
Moles, 130–132
Mountain Ash, 89
Mowers, 178–179
Mulch layers, 180
Mulches, 22, 33–34
Mustard greens, 39, 55, 71, 137, 149, 153

N
National Oceanic and Atmospheric Administration (NOAA) reports, 12
Native berries and fruits
    California blackberry, 102
    chokecherry, 103
    elderberry, 102–103
    sand cherry, 103
    serviceberries, 103
    Sierra plum, 103–104
    thimbleberry, 104

INDEX

western raspberry, 104
wild currant, 104
wild gooseberries, 104
wild grapes, 104
wild strawberries, 104
woods rose (rose hips), 105
Native flora, 3–4, 9, 17, 19, 143, 150
    propagation of, 157
Native soils, 2
Natural Resource Conservation Service (NRCS), 74
Newspaper mulches, 33–34
Nitrogen, 23, 25, 27, 29, 33
Nutrient levels, 22, 23–24
Nuts, 73
    acorns, 90
    almonds, 90
    chestnuts, 91
    pine nuts, 91–92
    walnuts, 92

O

Oak root fungus, 185
Onions
    direct seeding, 40, 55, 67
    fertilizing with manure, 27–28
    green, 57, 111, 194
    growing in heated greenhouse, 157
    growing in the hoop house, 154
    growing outdoors, 65, 148
    nutritional needs of, 24
    starting in the greenhouse, 57
    successive plantings of, 57
    when to plant, 56–57, 67, 68
Open pollination, 54
Organic matter, 21–22
Organic practices, 17, 19–20, 53
Outdoor planting, 148–151
Overhead watering, 46
Overwatering, 142

P

Paper wasps, 140
Parsnips, 24, 48, 55–57, 65–66, 190
Peaches, 78, 80
    suggested varieties, 79t
Pears, 80
  suggested varieties, 81t

Peas, 31, 39, 46, 55, 56, 61–62, 86, 151, 152
    insect pests, 137, 138
Peppers
    cold tolerance of, 11, 14, 15
    growing and harvesting, 61, 63–64, 93
    growing in heated greenhouses, 15, 156
    growing in hoop houses, 39, 125, 153
    as insect repellent, 130, 132
    nutritional requirements of, 24
    transplanting, 41, 43, 55, 122
    watering, 46
    when to plant, 56–57
Perennials, 19, 29, 53, 125, 150
Pesticides, 139
Pests
    common insect, 135t
    natural repellents for, 129, 130–134
    wildlife, 129–135
    See also Insects
pH levels
    of soil, 22, 23, 31, 185
    of water, 23
Phosphorous, 23, 28, 29, 32
Phytophora disease, 185
Picks, 50
Pine nuts, 91–92
Planet Jr. drop seeder, xv, 50, 58, 65, 121, 186, 189, 191, 192
Plant communities. See Flora and plant communities
Planting by the moon, 56
Planting time, 56–57. See also Planting schedules
Planting zones, 9–11, 140
PlantSkydd, 130
Plastic mulches, 34
Plows, 176, 176
Plums, 80
    Sierra, 103–104
    suggested varieties, 82t
Post hole diggers, 51
Potassium, 23, 29
Potatoes, 32, 33, 67, 150, 154
Precipitation, 2–3, 6–7
    average, 16t
    snowfall, 1, 3, 6–7
    thunderstorms, 13, 16

Prickly Pear Cactus, 89
Pruners, 49
Psyllid wasps, 140
Pumpkins, 41, 46, 55, 56, 57, 61, 69, 151

Q
Quince, 90
Quinoa, 92, 149

R
Rabbits, 130–131
Raccoons, 132
Radishes
 direct seeding, 40, 42, 55, 69, 186, 190, 192
 germination, 56, 192, 193
 growing outside, 14, 65, 149
 in hoop houses, 13, 15, 153
 irrigating, 43
 nutritional requirements of, 24
 in raised beds, 35, 191
 rotating, 194
 successive plantings of, 15, 57, 116, 124, 186
Raised beds, 35–36, 38, 38–39
 cover cropping in, 31
Rakes, 48
Raspberries, 95–96
 suggested varieties, 97t
 on trellises, 97
 western, 104
Reel mowers, 179
Resource Conservation District (RCD), 74
Rhubarb, 67
Ring rollers, 176. See also Rollers
Rippers, 177–178
Rock phosphate, 30
Rodents, 130–132
Rollers, 176, 177
Romanesco, 61
Romano, Gary
 family and background, xiii–xiv, xvi
 with Planet Jr. drop seeder, xv
Romano, Lou
 disking the soil, 160
 with potatoes, 32
Root crops. See Vegetable root crops
Rose hips, 105

Row covers, 123, 125, 125, 126

S
Sales and marketing, 172
Salinity, 22
Sand cherries, 103
Sandy loam, 21
Saws, 49
Scab, 33
Scallions, 148, 157
Season extension structures
 crop planning for, 124–125
 establishing seedbeds in, 120–123
 frost protection in, 123–124
 temperature management in, 119–120
 water and fertility in, 120
 winter crops in, 126–127
 See also Cold Frames; Geodesic domes; Greenhouses; Hoop houses
Seasons, 9
 fall, 14–15
 spring, 12–13
 summer, 13–14
 winter, 15–16
Seaweed emulsions, 29
Seed Savers Exchange, 54
Seed saving, 69–72
Seed screening, 70, 71
Seed spreaders, 49–51
Seed storage, 54, 70
Seedbeds
 preparation of, 41–42, 184, 185–189, 187, 188
 in season extension structures, 120–121
Seeders, xv, 50. See also Planet Jr. drop seeder
Seeding. See Direct seeding
Seeds
 heirloom, 53–54
 quinoa, 92
 sunflowers, 92–93
Serviceberries, 3, 183
Shallots, 65, 67, 109, 111, 194
Shock period, 41, 43, 54, 55
Shovels, 47
Siberian kale, 126
Sickle bar mowers, 179

INDEX

Sierra House Growing Dome Project, 116, 117
Sierra Nevada range, 1–2
    alpine region, 4
    climate, 5–7
    exposure, 4–5
    flora and plant communities, 3–4
    lower montane, 3
    native soils, 2
    precipitation and water systems, 2–3
    upper montane, 4
Sierra plums, 103–104
Sierra region, xvi–xix, 1. See also East slope
Sierra rose, 105
Sierra Valley Farms
    carrots at, 66
    field preparation, 184
    furrow row with transplants, 122
    greenhouse at, 112
    hoop houses at, 108, 112, 115, 191
    leafy greens harvest, 59
    map, xiii
    photo with outbuildings, 162
    raised beds at, 38
    row covers at, 125
    seeding greens at, 121
    solarization at, 141
    T-tape irrigation system at, 44–45, 121
    unworked land, 182
Silt loam, 21
Site selection, 35–39
    for fruit trees, 73–74
Skunks, 132
Snowfall, 1, 3, 6–7
Soil, 21–22
    and irrigation, 43–44
    native, 2
    pH of, 22, 23, 31, 185
    and seedbed preparation, 41
Soil Conservation Service soil report, 10–11
Soil erosion, 33
Soil temperature, 56
Soil testing, 35, 181, 185
Solarization, 140–141, 141
Sorrel, 149, 153
Spider mites, 137, 140
Spiders, 139
Spinach
    direct seeding, 40–42, 50, 55, 186, 190, 192
    germination, 56, 193
    insect pests, 136, 137
    growing at altitudes, 39
    growing and harvesting, 57–58
    growing in hoop houses, 13, 15, 153
    growing outdoors, 11, 14, 149
    growing in winter, 126
    nutritional requirements of, 24
    rotation planting, 124, 194
    site selection for, 35
    successive plantings of, 15, 57, 116
    transplants, 41
Sprayers, 181
Squash
    cold tolerance of, 11, 40
    cross-pollination, 69
    growing and harvesting, 61, 63
    growing in the greenhouse, 15
    growing in hoop houses, 39
    growing outside, 107, 151
    growing from seed, 41, 57
    insect pests, 136, 137, 138
    nutritional requirements of, 24
    in raised beds, 35
    transplanting, 55
    watering, 44, 46
    when to plant, 56
    See also Zucchini
Squash bugs, 138
Squirrels, 130–131
Stagger planting, 124–125. See also Succession planting
Strawberries, 101–102, 120, 129, 150, 153, 180
    wild, 104
Succession planting, 116, 148–150, 190–191. See also Stagger planting
Suggested varieties
    Apples, 76–78t
    Apricots, 79t
    Chestnuts, 91t
    Currants, 98t
    Figs, 88t
    Gooseberries, 99t

Grapes, 100t
Peaches, 79t
Pears, 81t
Plums, 82t
Walnuts, 92t
Sulfur, 24
Sunflowers, 46, 55, 92–93, 133, 149
Sunset Gardening Book, xvi, xvii
Swiss chard, 13, 55, 56, 57, 60, 124, 126, 148, 153, 186, 190, 192
    insect pests, 137

T

Temperatures, 6–7
    average, 16t
    in season extension structures, 119–120
    in spring, 12
    in summer, 13
    See also Climate; Frost-free days; Weather
Tent caterpillars, 139
Thimbleberries, 104
Thrips, 138
Thunderstorms, 13, 16
Thyme, 134
Tillers, 50, 177
Tomato hornworms, 138
Tomatoes
    in cold frames, 39
    determinate vs. indeterminate, 64, 125, 154, 155
    harvesting, 64
    in heated greenhouse, 156
    in hoop houses, 39, 64, 107, 115, 125, 155
    hybrids, 54
    insect pests, 136, 137, 138, 139
    irrigating, 43, 64, 180
    microclimates and, 40
    nutritional requirements of, 24
    outdoors, 151
    in raised beds, 35
    and root rot, 93
    soil temperatures for germination, 56
    transplanting, 41, 43, 54–55, 122
    as warm-season crops, 11, 12–15, 39, 41, 56, 57, 61
    water requirements of, 44, 46
Tool bar seeders, 179
Tools
    axes, 50
    common gardening tools, 47–51t
    cultivators, 50
    fertilizer/seed spreaders, 50–51
    forks, 48
    hoes, 48
    picks, 50
    post hole diggers, 51
    pruners, 50
    rakes, 48
    saws, 50
    for seed screening, 70
    seeders, 50
    shovels, 47
    tillers, 50
    See also Machinery and equipment
Tracklayers, 173. See also Tractors
Tractors
    drive train, 173
    mode of traction, 173
    photographs, 174, 175
    power size, 173
    selection of, 172–173, 175–176
    versatility, 174
Transplanters, 180, 180
Transplants
    photograph, 122
    raising your own, 43, 55, 57
    in season extension structures, 122–123
    vs. seeds, 40–41, 54–55
Traps, 131, 131, 132
Trees
    planting, 82–84, 83
    pruning, 84, 85
    watering, 84–86
    See also Fruit trees
Trenchers, 178
Trowels, 47
T-Tape irrigation, 42–45, 191–193
    photographs, 44–45, 121, 191
    in season extension structures, 121
Turnips, 40, 48, 55–57, 65–66, 71, 186, 190

INDEX

U
Upper montane, 4
USDA Hardiness Zones, 10–11, 40
Utility wheel tractors, 173. See also Tractors

V
Vegetable fruit crops
    artichokes, 60–61, 154
    corn, 11, 24, 35, 36, 41, 43, 46, 55, 56, 57, 61, 62, 107
    cucumbers, 14, 41, 46, 57, 61, 62, 125, 136, 152
    eggplant, 24, 39, 41, 43, 46, 55–57, 61, 62–63, 93, 107, 122
    melons and gourds, 63
    peas, 31, 39, 46, 55, 56, 61–62, 86, 137, 138, 151, 152
    pumpkins, 41, 46, 55, 56, 57, 61, 69, 151
    zucchini, 46, 55, 56, 61, 63, 69, 138, 144, 151
    See also Beans; Peppers; Squash; Tomatoes
Vegetable root crops, 11, 65, 121
    garlics, 24, 28, 56, 65, 67, 109, 111, 134, 136, 150, 154, 176
    horseradish, 67, 149
    Jerusalem artichokes, 28, 39, 65, 67
    leeks, 57, 109, 111, 148, 194
    parsnips, 24, 48, 55–57, 65–66, 190
    potatoes, 32, 33, 67, 150, 154
    rhubarb, 67
    scallions, 148, 157
    shallots, 65, 67, 109, 111, 194
    turnips, 40, 48, 55 57, 65–66, 71, 186, 190
    See also Beets; Carrots; Onions; Radishes
Vegetables
    asparagus, 60, 150, 157
    broccoli raab, 41, 61, 71, 156, 186, 190, 194
    Brussels sprouts, 24, 41, 44, 55, 57, 60, 61, 191
    celery, 24, 41, 55, 57, 154
    mizuna, 14, 15, 40, 42, 50, 55, 57, 58, 124, 149, 153
    mustard greens, 39, 55, 71, 137, 149, 153
    Romanesco, 61
    Swiss chard, 13, 55, 56, 57, 60, 124, 126, 137, 148, 153, 186, 190, 192
    See also Cabbage; Cauliflower; Greens; Kale; Lettuces; Spinach
Ventilation in hoop houses, 114–115, 115
Verticillium root rot, 93
Vole trap, 131
Voles, 130–131

W
Walnuts, 92
    Suggested varieties, 92t
Warm-season crops, 11, 35, 39, 41, 57
Wasps, 140
Water
    heating, 166
    in season extension structures, 120
    testing quality of, 22–23, 35
Water management, 166
Water systems, 2–3, 164–165
Watering, 164–165
    and fertilization, 30
    frequency of, 43–44, 46
    overhead, 46
    overwatering, 142
    See also Irrigation; Microsprinklers
Weather
    annual statistics, 17
    snowfall, 6–7
    thunderstorms, 13, 16
    weather station, 16
    See also Climate; Temperatures
Weed eaters, 179
Weeding, 194
Weeds, 140–142
    flaming, 141
    and fruit trees, 86–87
Well water, 164–165
West slope
    climate, 5–7
    crop selection, 39–40
    exposure, 5
    hardiness zones, 10

precipitation, 2–3
   See also Foothill region
Western raspberry, 104
What Grows Here in Plumas, Sierra, and Lassen Counties (Young), 10–11
White flies, 138, 140
Why I Farm: Risking It All for a Life on the Land (Romano), xiii–xvi, 163
Wild currants, 104
Wild gooseberries, 104
Wild grapes, 104
Wild strawberries, 104
Wildlife, 9, 17, 19
   bears, 134–135
   and berries, 94
   birds, 133
   deer, 133–134
   and fruit trees, 86, 87
   habitats for, 9, 19, 93, 143
   natural repellents, 129, 130–134
   raccoons, 132
   rodents, 130–131
   skunks, 132
Wind, 13, 16
Windbreaks, 74
Winter crops, in season extension structures, 126
The Winter Harvest Handbook (Coleman), 62, 109, 121
Wintergreens, 121
Won, SunMie, 37–38
Woods rose (rose hips), 105

Z

Zucchini, 46, 55, 56, 61, 63, 69, 144, 151
   insect pests, 138
   See also Squash

Thank you for purchasing a Bona Fide Book.
Please send any comments or suggestions
to editor@bonafidebooks.com.

Bang Printing is a member of the Green Press Initiative.
July & Winter was printed on SFI-certified paper.